Photosynthesis
energy transduction

a practical approach

KV-499-999

TITLES PUBLISHED IN

THE PRACTICAL APPROACH SERIES

Gel electrophoresis of proteins

Gel electrophoresis of nucleic acids

Iodinated density-gradient media

Centrifugation (2nd Edition)

Transcription and translation

Oligonucleotide synthesis

Microcomputers in biology

Mutagenicity testing

Affinity chromatography

Virology

DNA cloning

Plant cell culture

Immobilised cells and enzymes

Nucleic acid hybridisation

Animal cell culture

Human genetic diseases

Human cytogenetics

Photosynthesis energy transduction

a practical approach

Edited by

M F Hipkins

Department of Botany, University of Glasgow,
Glasgow, UK

N R Baker

Department of Biology, University of Essex,
Colchester, Essex, UK

IRL PRESS

Oxford · Washington DC

IRL Press Limited,
P.O. Box 1,
Eynsham,
Oxford OX8 1JJ,
England

©1986 IRL Press Limited

RESPECT COPYRIGHT
BERNE CONVENTION 100 YEARS
1886-1986
ENCOURAGE CREATIVITY

All rights reserved by the publisher. No part of this book may be reproduced or transmitted in any form by any means, electronic or mechanical, including photocopying, recording or any information storage and retrieval system, without permission in writing from the publisher.

British Library Cataloguing in Publication Data

Photosynthesis: energy transduction: a practical approach. — (Practical approach series)
1. Photosynthesis 2. Energy transfer
I. Hipkins, M.F. II. Baker, N.R. III. Series
581.1′3342 QK882

ISBN 0-947946-63-2 (hardbound)
ISBN 0-947946-51-9 (softbound)

PLYMOUTH POLYTECHNIC LIBRARY
Accn No 208975
Class No 581.13342 PHO
Contl No 0947946632

Printed by Information Printing Ltd, Oxford, England.

Preface

Photosynthesis — the conversion of light energy to chemical energy by plants and photosynthetic bacteria — comprises many diverse physical and chemical processes, ranging from the capture of light energy by the photosynthetic pigments to the chemical reactions of carbon assimilation. It is not possible in a book of this size to review all of the techniques required to investigate photosynthesis, so we have focussed our attention on a small number of processes which involve the first steps of energy conversion. We define energy transduction as those photosynthetic reactions leading to the reduction of NADP and to the photophosphorylation of ADP to ATP. But we do not attempt to cover the extremely specialised techniques needed to study the very rapid photophysical events of photosynthesis. This book, therefore, does not purport to be an all-encompassing review of photosynthetic methods.

Most of the techniques described in the individual chapters span from the relatively simple, using commercially available equipment, or equipment that can be built readily, to much more advanced techniques, where specialised equipment is essential. The simpler techniques are covered in some depth, but in the case of the more advanced techniques the coverage is more restricted, since they will inevitably have narrower applicability. In contrast to many of the previous volumes in this series, a number of chapters in the present volume contain detailed theoretical considerations of the principles underlying the methods.

In some instances, authors have specified particular suppliers of equipment. We would like to emphasise that in most cases equally acceptable alternatives exist, and specifying one supplier is not necessarily meant to exclude any others.

Finally, we thank the authors for their sterling efforts in producing their chapters, and for making accessible some of the techniques in photosynthesis that had formerly been the prerogative of the specialist.

M.F.Hipkins
N.R.Baker

Contributors

J.F.Allen
Department of Plant Sciences, Baines Wing, University of Leeds, Leeds LS2 9JT, UK

N.R.Baker
Department of Biology, University of Essex, Colchester CO4 3SQ, UK

M.F.Hipkins
Department of Botany, University of Glasgow, Glasgow G12 8QQ, UK

N.G.Holmes
Department of Plant Sciences, Baines Wing, University of Leeds, Leeds LS2 9JT, UK

R.C.Leegood
Research Institute for Photosynthesis, Department of Botany, University of Sheffield, Sheffield S10 2TN, UK

R.Malkin
Department of Molecular Plant Biology, University of California, 313 Hilgard Hall, Berkeley, CA 94730, USA

J.P.Markwell
Department of Agricultural Biochemistry, University of Nebraska, Lincoln, NE 68583, USA

J.D.Mills
Department of Biological Sciences, University of Keele, Keele, Staffordshire ST5 5BG, UK

Contents

Abbreviations

AM	assay medium
BSA	bovine serum albumin
CCCP	carbonyl cyanide *m*-chlorophenyl hydrazone
DAD	diaminodurene
DBMIB	2,5-dibromo-3-methyl-6-isopropyl-*p*-benzoquinone
DCCD	dicyclohexyl carbodiimide
DCMU	3-(3,4-dichlorophenyl)-1,1-dimethylurea
DCPIP	dichlorophenol indophenol
DMSO	dimethylsulphoxide
DPC	diphenyl carbazide
DTT	dithiothreitol
EDTA	ethylenediamine tetraacetic acid
e.p.r.	electron paramagnetic resonance
FCCP	carbonyl cyanide *p*-trifluromethoxyphenylhydrazone
FP	free pigment
Hepes	*N*-2-hydroxyethylpiperazine-*N*′-2-ethanesulphonic acid
IRGA	infra red gas analysis
LED	light-emitting diode
LHC	light-harvesting complex
LM	lysis medium
MARS	Markwell and Reinman System
MES	2-(*N*-morpholino)ethanesulphonic acid
MOPS	3-(N-morpholino)propanesulphonic acid
MV	methyl viologen
NADP	nicotinamide adenine dinucleotide phosphate
OD	optical density
PAGE	polyacrylamide gel electrophoresis
PAR	photosynthetically active radiation
PES	N-ethyl phenazonium ethosulphate
PD	*p*-phenylenediamine
Phaeo	phaeophytin
PMS	phenazine methosulphate
PPC	pigment-protein complex
PSI	photosystem I
PSII	photosystem II
RC	resistor-capacitor
RM	resuspension medium
SDS	sodium dodecylsulphate
T	percentage transmission
TEMED	N,N,N′,N′-tetramethylethylenediamine
TMPD	tetramethyl-*p*-phenylene diamine

CHAPTER 1

Introduction to Photosynthetic Energy Transduction

M.F. HIPKINS

1. DEFINITION

Photosynthesis is the conversion of light energy to chemical energy by plants and certain bacteria. Conceptually, photosynthesis can be divided into two phases. The first involves the capture of light energy and its transduction to chemical energy (ATP) and reducing equivalents (reduced nicotinamide adenine dinucleotide phosphate; NADP): these processes have often been called the light reactions. Secondly, the ATP and reduced NADP are consumed in the reduction of carbon dioxide to sugars by the carbon-assimilation cycle. In this book, photosynthetic energy transduction will be defined essentially as the light reactions, a definition which is certainly arbitrary, but the division of photosynthesis into two such phases is convenient, and has frequently been made.

In general the methods described in the following chapters are confined to higher plants, although many of them are directly applicable to algae, and some to photosynthetic bacteria.

This introductory chapter is intended only to enable the reader to put the techniques and methods described in other chapters into the context of the present concepts of the photosynthetic processes. It is therefore a brief survey of the salient points: the reader is directed to one of the several comprehensive surveys of photosynthesis for more detailed information (1 – 10).

2. THE PHOTOSYNTHETIC APPARATUS

In eukaryotes, photosynthesis takes place in organelles called chloroplasts: photosynthesis may be studied using a range of isolated preparations, from complex systems like protoplasts to the simpler chloroplast membrane components. The isolation of chloroplasts and their fractionation into functional components is described in Chapter 2. If the cellulose cell wall of a higher plant cell is digested with appropriate enzymes, a protoplast is liberated. Protoplasts are bounded by the cell membrane (the plasmalemma), and are frequently found to be an excellent way of preparing intact chloroplasts, by lysis of the cell membrane (Chapter 2). Intact chloroplasts retain their bounding envelope, which consists of a double membrane (*Figure 1*), and the enzymes of the carbon-assimilation cycle: they have the capacity to show bicarbonate-dependent light-induced oxygen evolution. The light reactions of photosynthesis occur in or on a complex network of thylakoid membranes found inside the chloroplast (*Figure 1*). Thylakoid membranes cannot assimilate carbon dioxide, but will show high rates of light-dependent

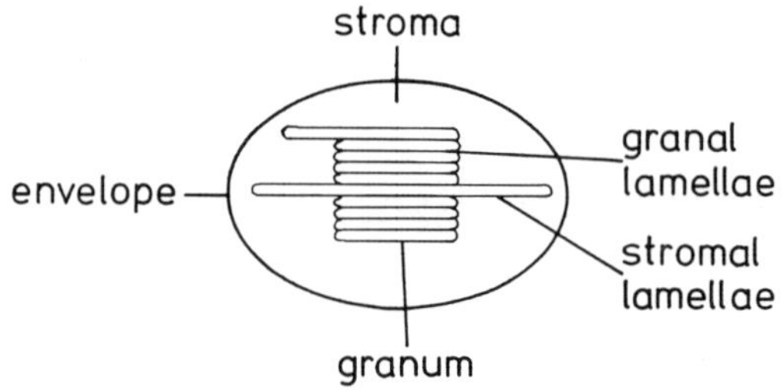

Figure 1. Diagrammatic representation of a mature higher-plant chloroplast. The chloroplast envelope (a double membrane) encloses the stroma, a matrix including all the enzymes of the carbon-assimilation cycle. The granal and stromal lamellae are formed into thylakoid vesicles, which are stacked in places to form grana.

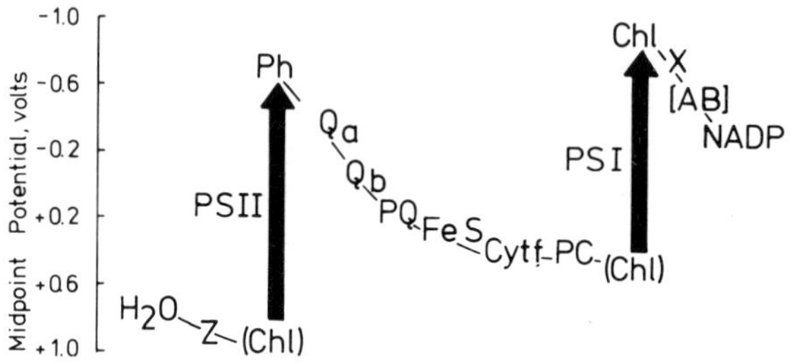

Figure 2. The 'Z-scheme' for non-cyclic electron transport in green plants, indicating the route taken by electrons from the oxidation of water to the reduction of NADP. The Z-scheme shows that the two photosystems are in series; the intermediates are arranged in order of mid-point redox potential, with strongly reducing couples at the top of the diagram. The reaction centre chlorophyll of photosystem II (PSII) is designated P680, that of PSI is designated P700. The components are: Z, unknown intermediate(s), probably involving Mn; Chl, chlorophyll; Ph, phaeophytin; PQ, plastoquinone; Q_a, Q_b, functionally distinct quinone molecules; FeS, the Rieske iron-sulphur centre (a non-haem iron); Cyt *f*, cytochrome *f*; PC, plastocyanin; X, A and B, iron-sulphur centres. For clarity, the scheme does not show some intermediates such as ferredoxin and Cyt b_{559}, nor does it incorporate Q- or b-cycles (*Figure 3*).

oxygen evolution when an exogenous electron acceptor is added (Chapter 5); they will also phosphorylate ADP to ATP (Chapter 6).

All the pigment-protein complexes and most of the components of the photosynthetic electron transport chain are intrinsic components of the thylakoid membrane. Some electron transport components are extrinsic to the thylakoid and are found on its surface. Disruption of the membrane, mainly by use of detergents, is the first step in the isolation of the membrane components (Chapter 2). *Figure 1* indicates the structural features of chloroplasts, and emphasises that the chloroplast contains two spaces: the intrathylakoid space (lumen) enclosed by the thylakoid membranes, and the stroma (containing the enzymes of carbon assimilation) which is enclosed by the chloroplast envelope.

3. PHOTOSYNTHETIC ELECTRON TRANSPORT

One of the requirements of the enzymic processes leading to the reduction of carbon dioxide is reducing equivalents: these are furnished by the photosynthetic electron transport chain (the 'Z-scheme', *Figure 2*), which serves to transfer electrons from water to the terminal electron acceptor NADP. This is an energy-requiring process since water is a relatively stable molecule, and reduced NADP is a strong reducing agent. The necessary energy is provided by light.

3.1 **Light Absorption by the Photosynthetic Pigments**

In green plants the photosynthetic pigments comprise chlorophyll *a*, chlorophyll *b* and several carotenoids, these being highly organised to make light harvesting efficient. Chlorophyll *b* and the carotenoids are collectively called accessory pigments, and pass the light energy they absorb to chlorophyll *a*, thus increasing the spectral range for photosynthetically active radiation. In addition, chlorophyll *a* molecules themselves are organised into 'photosynthetic units' where, on average, 250 – 300 chlorophyll *a* molecules capture light energy and pass it to a single specialised chlorophyll *a* molecule in a reaction centre, where the photochemical reactions take place.

3.2 **Reaction Centres**

Reaction centres are specific pigment-protein complexes which contain chlorophyll *a* molecules in a special environment (Chapter 3). In oxygen-evolving organisms there are two types of reaction centre, designated photosystem I and photosystem II. Within the reaction centres the critical photochemical reactions take place: light energy is used to promote the reaction centre chlorophyll P to an excited state P^*; this then reduces an electron acceptor (A), and hence P becomes oxidised. Re-reduction of P^+ takes place by oxidation of an electron donor D. In summary:

$$\begin{aligned} D\,P\,A + h\nu &\rightarrow D\,P^*A \\ D\,P^*A &\rightarrow D\,P^+A^- \\ D\,P^+A^- &\rightarrow D^+P\,A^- \end{aligned}$$

D^+ and A^- are then reduced and oxidised respectively by secondary electron transport. In photosystem I the reaction centre chlorophyll is designated P700, since when photo-oxidised to $P700^+$ a small bleaching (decrease in absorption) at approximately 700 nm is seen. Likewise the reaction centre chlorophyll of photosystem II is designated P680.

3.3 **The Z-scheme**

Figure 2 indicates that the two reaction centres of oxygen-evolving organisms are arranged in series. The terminal electron acceptor, NADP, is reduced by electrons from photosystem I; these electrons originate from the oxidation of water by photosystem II. The Z-scheme is drawn on a scale of redox potential, where those compounds with lower mid-point redox potentials (strong reducing agents) are found at the top of the diagram. The scheme is thus divided into the light-driven 'uphill' sections at the reaction centres where the energy of a photon is used to reduce an electron acceptor, and hence make a reducing agent, and the other 'downhill' sections which yield energy as electrons pass from stronger reducing agents to relatively weaker ones. The energy so released is subsequently used to phosphorylate ADP to ATP (Section 4).

The intermediates of the electron transport chain can be divided into several classes: cytochromes (cytochromes *f*, b_6 and b_{559}), non-haem iron-sulphur centres (e.g. ferredoxin and the Rieske centre), quinones (e.g. plastoquinone), proteins with transi-

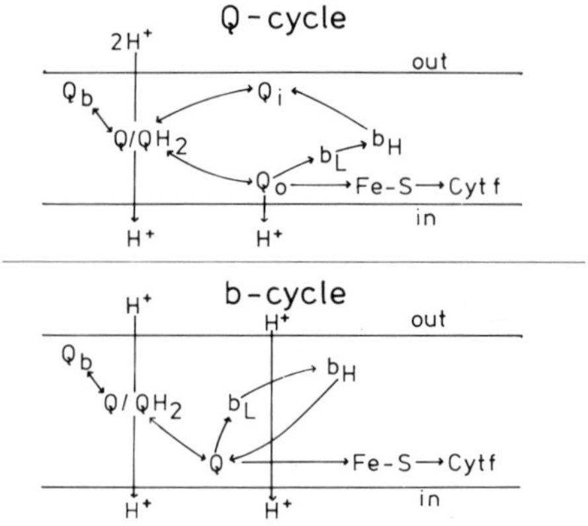

Figure 3. Two schemes (the Q- and b-cycles) for the involvement of the cytochrome b_6-f complex in electron transport and coupled proton movements across the thylakoid membrane.

tion metals as the prosthetic group (plastocyanin, Mn-containing proteins), chlorophyll and phaeophytin. Their relative positions in the electron transport chain are shown in *Figure 2*. The biochemical nature of the components of the electron transport chain is quite varied, but all have the property of being able to accept and pass on electrons, that is, to be reduced and subsequently oxidised. The components can, however, be split into two functional groups: those reduced by electrons only, and those reduced by electrons and protons. This distinction is important in considering the mechanism of photophosphorylation (Section 4; Chapter 6).

3.4 **Cyclic and Non-cyclic Electron Transport**

Figure 2 shows non-cyclic electron transport where the oxidation of water is linked both to the reduction of NADP and to photophosphorylation, and both photosystems participate. Another mode of electron transport, involving photosystem I only, also exists. In this so-called cyclic electron transport there is no net oxidation of electron donors or reduction of acceptors, but photophosphorylation still takes place. There has been a good deal of discussion recently about the involvement of the cytochrome b_6-f complex in both cyclic and non-cyclic electron transport: two possible schemes, called Q- and b-cycles are shown in *Figure 3*.

The rate of electron transport through the chain can be measured

(i) by detecting the evolution of oxygen when water is oxidised (Chapter 5),
(ii) by detecting the uptake of oxygen when an exogenous electron acceptor such as methyl viologen is reduced (Chapter 5), or
(iii) by detecting reduction of coloured exogenous electron acceptors spectrophotometrically (Chapters 4 and 5).

Moreover the presence of intermediates like cytochrome f and P700 can be assessed spectrophotometrically.

4. PHOTOPHOSPHORYLATION

The carbon-assimilation cycle not only requires reducing equivalents in the form of

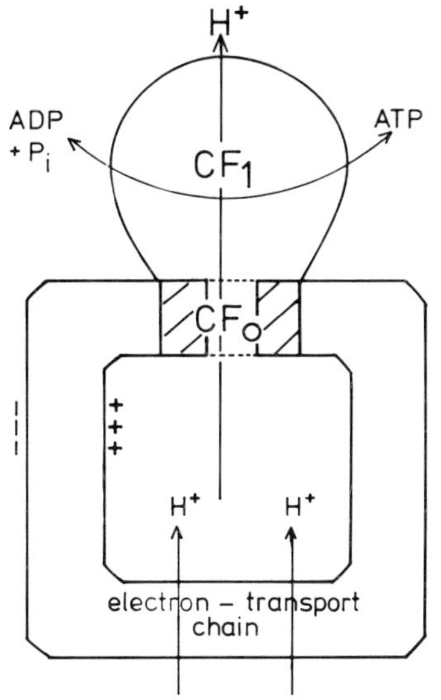

Figure 4. A schematic diagram showing the main postulates of the chemiosmotic hypothesis for phosphorylation. Phosphorylation involves a closed vesicle formed from a membrane that is essentially H^+-impermeable. Light-induced electron transfer leads to the build-up of an electrochemical potential gradient of protons across the membrane, comprising a pH-gradient and an electrical potential gradient. ATP is formed when the protons move down their electrochemical potential gradient through a vectorial, membrane-bound ATPase.

reduced NADP, but also chemical energy in the form of ATP. The energy for phosphorylation is derived from electron transport (see above); most workers now agree that the chemiosmotic hypothesis is a good working model for this process. The hypothesis has four main postulates, here applied to the chloroplast (*Figure 4*):

(i) phosphorylation is associated only with closed vesicles whose membranes are essentially impermeable to protons — in the case of chloroplasts the thylakoid vesicles and membranes,

(ii) the high-energy state or energetic intermediate between electron transport and phosphorylation is a transmembrane electrochemical gradient of protons (the protonmotive force), comprising a proton concentration gradient, ΔpH, and an electrical potential gradient, $\Delta\psi$,

(iii) the protonmotive force is created by the translocation of protons across the thylakoid membrane concomitant with either cyclic or non-cyclic electron transport,

(iv) phosphorylation is linked to the efflux of protons from the intrathylakoid space through a specific, vectorial H^+-translocating ATPase.

The evidence for the chemiosmotic hypothesis is now good, although there is still debate about whether there is a bulk electrochemical potential gradient between the thylakoid lumen and the stroma; in addition to the precise mechanism of phosphorylation by the ATPase is unresolved.

The second postulate noted above is central to the chemiosmotic hypothesis since it predicts that the electron transport chain should translocate protons. Light-induced H^+ uptake by thylakoid membranes is well documented, and results from the alterna-

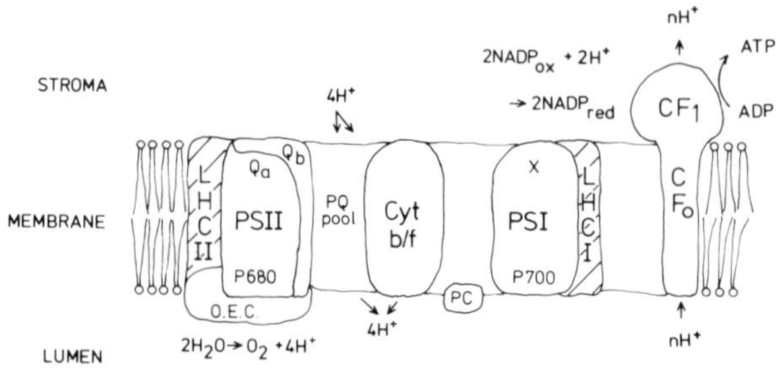

Figure 5. Diagram indicating how non-cyclic electron transfer is thought to be coupled to proton movements across the thylakoid membrane. (a) The oxidation of water takes place in the lumen, leading to the release of protons, whilst proton uptake occurs in the stroma when NADP is reduced. (b) The reduction of quinones, which requires protons as well as electrons, takes place at the stromal side of the membrane, whilst oxidation of plastoquinone and proton release occurs on the lumen side, leading to proton translocation. See also *Figure 3*.

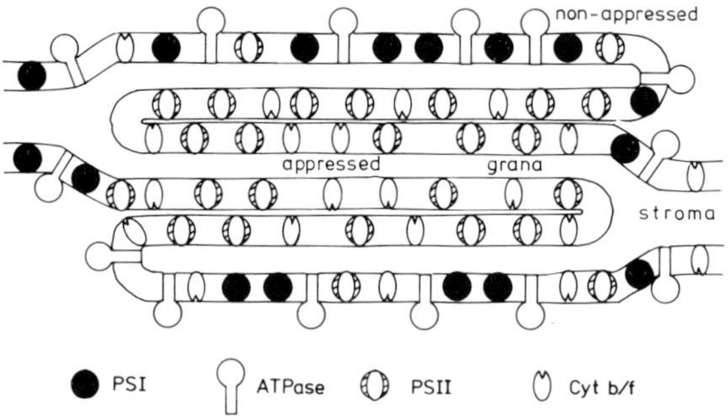

Figure 6. A hypothetical scheme for the lateral distribution of membrane complexes between the appressed and non-appressed regions of the thylakoid membranes. Photosystem II is largely confined to the appressed regions, whilst photosystem I and the ATPase are restricted to the non-appressed regions. In this scheme the cytochrome *b-f* complex is distributed evenly between the two regions; other models place the cytochrome complex largely in the unappressed regions.

tion of electron-carrying and hydrogen-carrying components in the Z-scheme and from these carriers being arranged vectorially across the membrane (*Figure 5*). Essentially, proton translocation arises from the reduction of NADP at the stromal side of the thylakoid membrane with the consumption of protons, and the oxidation of water taking place in the lumen, leading to the liberation of protons. In addition, plastoquinone requires both protons and electrons to be reduced: it is reduced at the stromal side and oxidised at the lumen side, again leading to proton translocation across the membrane. However, the nature and role of any Q- or b-cycle in proton pumping (*Figure 3*) is still under discussion.

5. THE COMPLEXES OF THE THYLAKOID MEMBRANE

The number and nature of the pigment-protein and other complexes of the thylakoid membrane (*Figure 6*) is a very active field of research. Particularly in the area of the light-harvesting complexes associated with the photosystems, the picture is changing rapidly (Chapter 3).

5.1 Sub-chloroplast Fractions

Stromal thylakoid membranes contain rather little photosystem II, whilst granal membranes contain predominantly photosystem II. Separation of the two classes of membranes by detergent treatment can yield fractions enriched in either photosystem (Chapter 2), but which still contain a large amount of light-harvesting pigments.

5.2 Reaction Centre Preparations

Disruption of the respective sub-chloroplast fractions by further detergent treatment can decrease the amount of light-harvesting pigment, but in higher plants there appears to be a 'core complex' of light-harvesting pigment attached to both types of reaction centre, and which is very difficult to remove (Chapters 2 and 3).

5.3 Light-harvesting Pigment-protein Complexes

The nature and number of light-harvesting complexes is perhaps the least resolved aspect of the thylakoid membrane components (Chapter 3). It seems possible that there are three types of light-harvesting complexes: firstly, the pigment-protein complexes which are tightly bound to the reaction centres; secondly, the pigment-protein complexes less tightly bound, and which may be removed by certain procedures, and thirdly the 25-kd light-harvesting pigment-protein complex that can be detected easily using polyacrylamide gel electrophoresis (PAGE) analysis (Chapter 3). This last complex was thought to have a role in regulating the relative distribution of light energy between photosystems I and II.

5.4 The Cytochrome b_6-f Complex

This complex has no light-harvesting or photochemical function, but participates in electron transport from photosystem II to photosystem I, and in cyclic electron flow around photosystem I. The complex has received a good deal of attention recently since it is thought to participate in a Q- or b-cycle which leads to the build-up of an electrochemical potential gradient of protons across the thylakoid membrane.

5.5 The Coupling Factor Complex

Photophosphorylation involves the vectorial movement of protons down their electrochemical potential gradient from the lumen of the thylakoid to the stroma *via* the coupling factor (Chapter 6). The energy yielded by the protons is taken up in phosphorylating ADP to ATP. The complex consists of two parts: CF_0 is an intrinsic membrane protein which acts as a proton-conducting channel; CF_1 is an extrinsic protein attached to CF_0 (Chapter 2), and which has both catalytic and regulatory sites (Chapter 6).

The spatial distribution of these complexes in the thylakoid membrane is not uniform. Photosystem I and the coupling factor are predominantly found in the non-appressed regions of the membrane (i.e. at the periphery of the granal stacks and on the stromal lamellae) whilst photosystem II is found in the appressed regions (*Figure 6*). The location of the cytochrome b_6-*f* complex is under discussion, but *Figure 6* shows one possibility where it is distributed between the appressed and non-appressed regions.

6. ACKNOWLEDGEMENTS

I thank Christine Raines for drawing the figures.

7. REFERENCES

1. Hipkins,M.F. (1984) in *Advanced Plant Physiology*, Wilkins,M.B. (ed.), Pitman, London. p. 219.
2. Hall,D.O. and Rao,K.K. (1981) *Photosynthesis*, 3rd edition, published by Edward Arnold, London.
3. Danks,S.M., Evans,E.H. and Whittaker,P.A. (1983) *Photosynthetic Systems: Structure, Function and Assembly*, published by John Wiley, Chichester.
4. Prebble,J.N. (1981) *Mitochondria Chloroplasts and Bacterial Membranes*, published by Longman, London.
5. Clayton,R.K. (1980) *Photosynthesis: Physical Mechanisms and Chemical Patterns*, published by Cambridge University Press.
6. Nicholls,D.G. (1982) *Bioenergetics: An Introduction to the Chemiosmotic Theory*, published by Academic Press, London.
7. Edwards,G. and Walker,D.A. (1983) *C_3, C_4: Mechanisms, and Cellular and Environmental Regulation, of Photosynthesis*, published by Blackwell, Oxford.
8. Govindjee, ed. (1982) *Photosynthesis*, 2 Vols, published by Academic Press, London.
9. Barber,J. ed. (1976) *Topics in Photosynthesis*, Vols **1-6**, published by Elsevier, Amsterdam.
10. Halliwell,B. (1981) *Chloroplast Metabolism: The Structure and Function of Chloroplasts in Green Leaf Cells*, published by Oxford University Press.

CHAPTER 2

Isolation of Sub-cellular Photosynthetic Systems

R.C. LEEGOOD and R. MALKIN

1. INTRODUCTION

This chapter describes standard procedures for the isolation of intact chloroplasts (both mechanically and from protoplasts), thylakoids and membrane complexes.

The photosynthetically active fractions whose isolation is described in this chapter (*Figure 1*) have a range of photosynthetic competence. Intact chloroplasts retain their bounding envelope and can, to varying degrees, fix carbon dioxide. After lysis of the chloroplast envelope the thylakoid membranes can be isolated. If provided with an exogenous electron acceptor, thylakoid membranes can catalyse the photooxidation of water and, provided the necessary substrates and cofactors are present, this electron transport can be coupled to phosphorylation. Thylakoid membranes can be treated with detergent, and sub-chloroplast fractions which show the characteristics of photosystem II (PSII) or photosystem I (PSI) may be separated. These fractions can be treated further to yield their respective reaction centre complexes. Other fractionation procedures can be employed to yield complexes which are not in themselves photoactive, but which participate in electron transfer (the cytochrome b_6-f complex) or photophosphorylation (the ATP synthetase) (*Figure 1*).

The isolation of chloroplasts which retain their outer envelope is obviously a prerequisite for the maintenance of any process whose activity is located partly or wholly in the stroma, such as CO_2 fixation, the metabolism of nitrogen, sulphur and lipids, protein synthesis and the uptake of substances across the chloroplast envelope. The envelope of the intact chloroplast also provides a protective jacket which permits the isolation of the thylakoids, stromal enzymes and genetic material within, free from mechanical damage, the action of proteases, phenolics etc., and which prevents the loss of loosely-bound proteins from the thylakoids. For example, thylakoids isolated direct from intact chloroplasts show higher P/O ratios and better photosynthetic control (Chapter 5) than those isolated direct from the leaf material. Another advantage is that where stromal enzymes or other components have counterparts in other cellular compartments, isolation of intact chloroplasts constitutes a simple and highly efficient preliminary purification. However, it should not be assumed that because a chloroplast is intact its components are necessarily fully active. It is important for many purposes to isolate material which can catalyse its various activities at physiological rates. CO_2 fixation is certainly very sensitive to inactivation during the isolation process and can be measured very rapidly, so it can usefully be employed as a general indicator of the metabolic competence of isolated chloroplasts.

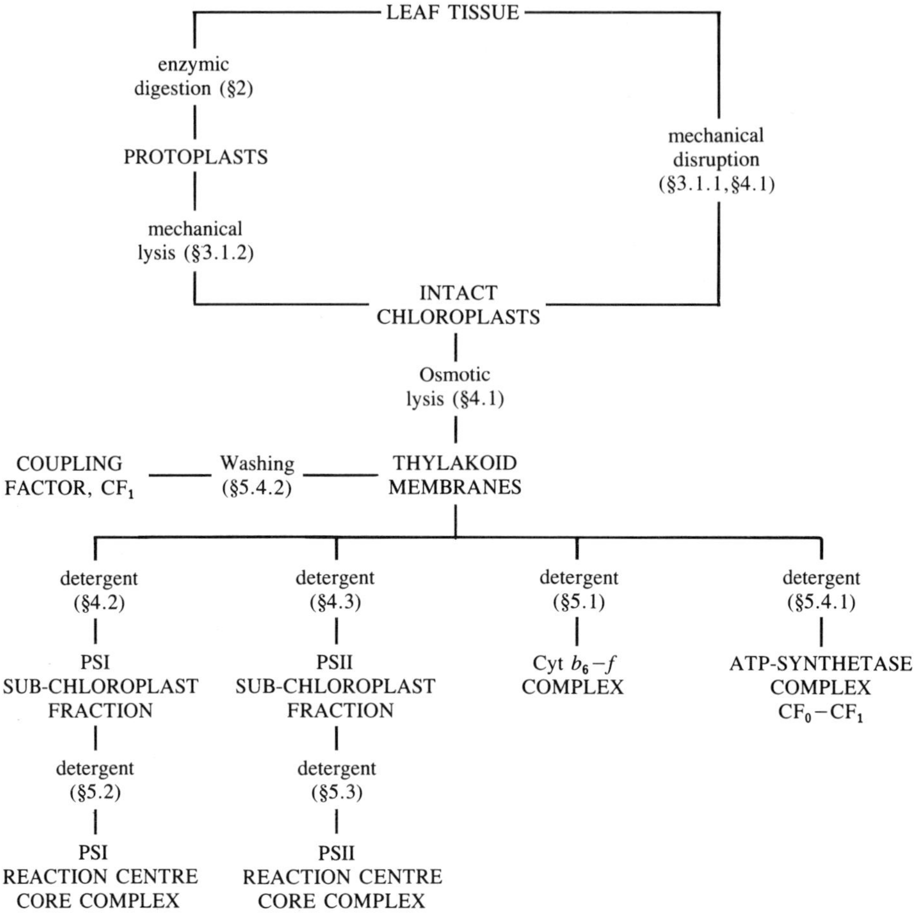

Figure 1. Flow diagram indicating the steps involved in isolating sub-cellular photosynthetic preparations. The sections in the chapter where methods are described are indicated.

1.1 Plant Material

It cannot be over-emphasised that poor starting material will not yield good chloroplasts. Leaves of spinach (*Spinacea oleracea*, other species called spinach yield indifferent chloroplasts) must be low in starch and calcium oxalate (both recognised by a white ring in the chloroplast pellet after centrifugation) and should therefore be taken from actively growing, non-flowering plants. To achieve this, plants are ideally grown in short days (~ 8.5 h) and require bright light. A suitable variety is US National Hybrid 74 (Ferry Morse Seed Co., P.O. Box 100, Mountain View, CA 94042, USA), sold in the UK and Australia as Yates Hybrid 102.

Peas are a useful alternative to spinach. Growth should yield uniform, actively-growing seedlings of about 5 cm in height after 9 – 11 days (see, for example, 1 – 4). Dwarf varieties (e.g. Feltham First, Little Marvel) give good results. Protoplasts can be prepared from older leaves of grasses, but maximum yields are obtained from wheat, barley and maize plants which are between 1 and 2 weeks old. Even so, not all varieties are ideal, so a range of varieties should be tested first. Mature leaves of spinach and peas are suitable for protoplast preparation, although their susceptibility to digestion depends on growth conditions and season. When isolating sub-chloroplast fractions, either spinach or pea is normally used as starting material.

2. ISOLATION OF PROTOPLASTS

Protoplasts capable of fixing CO_2 at high rates have been prepared from a wide range of species. The method of preparation involves exposure of the leaf cells to a mixture of cell wall-degrading enzymes for a few hours, followed by purification of the isolated protoplasts, usually by flotation in a gradient (5). Apart from being used for the isolation of highly intact, highly active chloroplasts, protoplasts have also been used for the preparation of other organelles such as mitochondria and vacuoles (6), in metabolic studies (although for work on photosynthesis, the absence of a sink probably distorts metabolism), besides regeneration and genetic studies.

With plants which perform C_4 photosynthesis (e.g. maize) bundle-sheath cells are usually isolated (7,8) together with mesophyll protoplasts, as protoplasts cannot be isolated in appreciable amounts from bundle-sheath tissue because of the nature of the bundle-sheath cell wall [although Edwards *et al.* (9) have had success in isolating bundle-sheath protoplasts from *Panicum miliaceum*, and Moore *et al.* (10) have had some success using the cellulase Onozuka RS (2.1) on C_4 dicotyledonous species such as *Atriplex spongiosa* and *Flaveria trinervia*]. Intact cells can also be isolated very easily from a number of plants. They can be prepared either by mechanical means in high yields from a few species (*Asparagus, Xanthium, Digitaria sanguinalis*) or by digestion with pectinase in the absence of cellulase (soybean, tobacco, *Vicia faba*, for references see 11). The main advantage is ease of preparation and ready applicability of cells as a test-system for herbicides etc. as well as for studies of whole cell photosynthesis. One important advantage of cell preparations from C_4 plants such as *D. sanguinalis* is that they are permeable to substrates such as 3-phosphoglycerate, whereas C_4 mesophyll protoplasts are, for practical purposes, impermeable to such substrates and their use is generally limited to the preparation of chloroplasts and to enzyme localisation studies.

2.1 Pectolytic and Cellulolytic Enzymes

There are now many commercial enzyme preparations available and a few examples are provided below.

(i) Cellulase from *Trichoderma viride*. Available as Onozuka RS or R10 (Yakult Honsha Co. Ltd., 8-21 Shingikancho, Nishinomiya, Japan) or as Cellulysin (Calbiochem). Onozuka RS is approximately twice the price of R10 but is of higher specific activity, has higher xylanase activity and dissolves cell walls from a wider range of material (see, for example, 10). Cellulase 3S (widely cited in the literature) is no longer available. Cellulase is usually employed at concentrations of between 1 and 3% (w/v), the concentration required should be determined for each tissue.

(ii) Pectinase from *Rhizopus* spp. As Macerozyme R-10 (Yakult) or Macerase (Calbiochem). Pectinase is used at concentrations of between 0.1 and 0.5% (w/v).

(iii) Pectinase from *Aspergillus* spp. As Rohament P (Röhm GmbH, Darmstadt, FRG) or as Pectolyase Y23 (Seishin Pharmaceutical Co. Ltd., Chiba-ken, Japan).

(iv) Driselase, a mixture of cellulase, pectinase and other enzymes from Basidiomycetes (Kyowa Hakko Kogyo Co. Ltd., 1-6-1, Ohtemachi, Chiyoda-Ku, Tokyo, Japan).

All these preparations are relatively expensive. However, Pectolyase Y23 can be purified very simply from a culture of *Aspergillus japonicus* grown on wheat bran (12). Used enzymes can also be recycled. The discarded enzyme medium is deep-frozen and once a large enough batch has accumulated, the enzymes may be concentrated by ammonium sulphate precipitation and freeze-dried for storage.

2.2 General Procedures for Protoplast Preparation

2.2.1 *Preparation of the Tissue*

The aim of tissue preparation is to expose mesophyll cell walls of leaves to the solution of digestive enzymes. This can be accomplished by a number of means, the simplest of which is to cut the leaf into thin segments between 0.5 and 1 mm wide. This works well for C_3 and C_4 grasses, but usually requires some practice. Some advantage (e.g. in the case of spinach) may be obtained by cutting the leaves into segments under 0.5 M sorbitol. The leaf strips are gently drained before being placed in the digestion medium. Vacuum infiltration may also be employed to facilitate penetration of the enzyme. Alternatively the epidermis of the leaf may be stripped (spinach, peas) or the surface of the leaf gently abraded using a toothbrush [sunflower (13)] or carborundum powder. The leaves are placed with the stripped or broken surface in contact with the enzyme medium.

2.2.2 *Incubation of the Tissue*

(i) Dissolve the pectinase and cellulase mixture in a medium containing an osmoticum (usually sorbitol), some divalent cations (which prevent clumping) and which is lightly buffered at pH 5.0 – 5.5. For wheat, barley, peas and spinach a suitable medium is 0.5 M sorbitol, 1 mM $CaCl_2$, 5 mM MES (pH 5.5) containing 2% cellulase and 0.3% pectinase. Sometimes bovine serum albumin (BSA) may be included and Mg^{2+} may be substituted for Ca^{2+} [e.g. in the preparation of maize mesophyll protoplasts (14)]. The enzymes usually take a few minutes to dissolve in medium I (*Table 1*) and can conveniently be left to dissolve in a refrigerator overnight.

(ii) Adjust the pH to pH 5.5 using dilute HCl.

(iii) Put the medium into a large shallow dish (50 ml in a 19 cm diameter crystallising dish), add the leaf strips or segments as they are prepared, and gently submerge them in the medium using the blade of a spatula. The tissue segments should form an even layer over the bottom of the dish; overcrowding will not improve the yield.

(iv) The incubation is normally done at 28 – 30°C without shaking. Lower temperatures have also been employed for longer periods (9); higher temperatures aid digestion but tend to yield poorer protoplasts. Illumination during the incubation results in higher yields and better photosynthetic performance; a 150 W

Table 1. Purification of Protoplasts.

1. Remove the incubation medium and discard it. Gently wash the protoplasts out of the tissue in 3 × 20 ml of ice-cold 0.5 M sorbitol, 1 mM $CaCl_2$, 5 mM MES (pH 6.0) (medium I). Pass the washings through a 500 μm mesh, followed by a 200 μm mesh.
2. Centrifuge at 100 *g* for 5 min and discard the supernatant by rapid inversion of the tube.
3. Resuspend the pellet, with very gentle shaking, in three drops of ice-cold 0.5 M sucrose, 1 mM $CaCl_2$, 5 mM MES (pH 6.0) (medium III). Add a further 5 ml of the sucrose medium and mix by gentle inversion.
4. Pipette 2 ml of 0.1 M sorbitol, 0.4 M sucrose, 1 mM $CaCl_2$, 5 mM MES (pH 6.0) (medium II) on top of the sucrose layer, followed by 1 ml of 0.5 M sorbitol, 1 mM $CaCl_2$, 5 mM MES (pH 6.0). Ensure that the layers do not mix by pipetting them in slowly.
5. Centrifuge at 250 *g* for 5 min and carefully collect the protoplasts (which float to the bottom of the uppermost layer) using a Pasteur pipette.

bulb suspended about 25 cm above the dish is sufficient. The incubation period is not usually more than 3 h, and may be less, particularly if cellulase RS and/or Pectolyase Y23 is employed (e.g. 10,12). Incubation periods of longer than 3 h, while they may result in higher yields, may also lead to deterioration of the resultant protoplasts.

2.2.3 *Protoplast Purification*

Many protoplasts are broken during the incubation and preliminary isolation so that the intact protoplasts are heavily contaminated by chloroplasts and by cell debris. However, intact protoplasts are less dense than chloroplasts and can be separated from them by flotation (*Table 1*). First, the protoplasts must be washed out of the tissue. If digestion has been particularly effective, some protoplasts may have been released into the medium, in which case the medium can be centrifuged and the protoplasts retrieved. The first mesh (500 μm − usually a tea-strainer) is designed to retain large pieces of tissue; these should be returned to the dish and washed out gently three times. Sometimes a little agitation of the tissue with a spatula will improve the yield. The 200 μm mesh (Henry Simon Ltd., Stockport, Cheshire) retains vascular strands. With C_4 plants, an 80 μm mesh is used to retain isolated bundle-sheath strands. For the centrifugation, any bench centrifuge will do. Simply determine a speed adequate to produce a compact pellet, yet not so compact as to be difficult to resuspend by gentle shaking. Glass or plastic centrifuge tubes are suitable, but polycarbonate tubes should be avoided.

After centrifugation in the gradient, intact protoplasts should appear as a band beneath medium I, all other debris being present in the pellet. Quite frequently appreciable numbers of protoplasts may remain in the pellet. This can usually be overcome either by making the whole gradient more dense (e.g. by increasing the concentrations of sucrose and sorbitol to 0.6 M) or by adding 5−10% dextran (15 000−20 000 mol. wt.) to increase the density of medium III. Sucrose and sorbitol may also be replaced by varying concentrations of Percoll (15).

Care should be taken when collecting the protoplasts, since they break very easily. Similarly, if the protoplasts are used for photosynthetic studies, they must be stirred very gently. The stability of protoplasts from C_3 species is normally greatest when they are stored on ice in medium I, and considerable photosynthetic activity may be retained for many hours after isolation (see 16,17).

Protoplast intactness can be rapidly assessed by microscopic observation. A quantitative estimate of protoplast intactness may also be made, using an oxygen electrode (Chapter 5), by measuring the activity of glycollate oxidase, an enzyme which is released from damaged protoplasts, before and after the addition of a detergent which disrupts the cell membrane (15).

3. ISOLATION OF CHLOROPLASTS

3.1 **Procedures**

There are two procedures for the isolation of chloroplasts. The first is the mechanical disruption of the tissue. Its advantages are its rapidity, its economy and high yield of stable chloroplasts. Its principal disadvantage is that its application is restricted to but a few species, principally spinach and peas, nearly all other species yielding chloroplasts which are either broken or which fix CO_2 at only poor rates, although in cases where good thylakoid membranes or stromal enzymes are required from intact chloroplasts, the loss of CO_2 fixation capacity is less likely to be of importance. For example, mechanically-prepared intact chloroplasts of sunflower are largely inactive in CO_2 fixation, but the membranes and stromal enzymes may be reconstituted and will fix CO_2 at high rates (18).

The second method of chloroplast preparation employs protoplasts. Its outstanding advantage is its applicability (or its potential applicability) to a far wider range of species than mechanical methods. Protoplasts provide a source of freshly prepared chloroplasts throughout the day. The principal disadvantages of protoplasts are the time-consuming preparation, the expense of cell wall-degrading enzymes and the relatively low yield of chloroplasts compared with mechanical procedures. Chloroplasts prepared from protoplasts also tend to be less stable (serious loss of the capacity to fix CO_2 during 1-h storage) than chloroplasts isolated mechanically, which are usually stable for a whole day.

3.1.1 *Mechanical Preparation of Chloroplasts*

The procedures for the mechanical preparation of chloroplasts are summarised in *Table 2*. All glassware should be thoroughly cleaned and free from detergent. The entire pro-

Table 2. Preparation of Spinach Chloroplasts.

1. Float the leaves on cold running water and illuminate them for 20 – 30 min, then remove the midribs and cut the leaves into strips.
2. Mix 80 g of leaves with 250 ml of semi-frozen grinding medium (0.33 M sorbitol, 10 mM $Na_4P_2O_7$, 5 mM $MgCl_2$, 2 mM sodium iso-ascorbate, freshly prepared and adjusted to pH 6.5 with HCl). Homogenise for 3 sec in a Polytron or 3 – 5 sec in a Waring blender.
3. Squeeze the brei through two layers of muslin to remove coarse material, then quickly filter the brei through a sandwich of cotton wool between eight layers of muslin (previously soaked in grinding medium).
4. The centrifugation should be rapid and should be done as soon as possible after homogenisation. Ideally a swing-out head should be accelerated to 6000 – 8000 *g* and returned to rest within 90 sec.
5. Discard the supernatant and wash the surface of the pellet by gently swirling with a few ml of grinding or resuspension medium. Resuspend the pellet in a small quantity (0.5 ml) of ice-cold resuspension medium [0.33 M sorbitol, 2 mM EDTA, 1 mM $MgCl_2$, 1 mM $MnCl_2$, 50 mM Hepes-KOH (pH 7.6)].

cedure should be carried out at 4°C. Pre-illumination of the tissue is not essential but results in shortened induction periods and higher rates. The media used for chloroplast isolation differ widely. For spinach, pyrophosphate is used as a buffer because it is both cheap and effective, but Good's buffers are equally effective and are employed in resuspension and assay media.

The use of a Polytron blender (Kinematica GmbH, Zurich, Switzerland) for homogenising the tissue is preferred, but most blenders can yield good results. The grinding medium can be converted into a semi-frozen slush by stirring it in a bath of alcohol kept in the deep-freeze. The Perspex container used with a Polytron is square in cross-section (5 × 5 × 20 cm high). The homogenisation should be brief and the filtration and centrifugation steps as speedy as possible. Nylon mesh may be substituted for muslin, but the latter is more efficient in retaining broken fragments of chloroplasts.

Resuspension of the chloroplast pellet can be done in a variety of ways, using a paint-brush, a piece of cotton wool attached to a glass rod or simply by gentle swirling or shaking. The chloroplast suspension is then stored on ice. Depending on the intactness achieved (which may often be 90% or more if good starting material is employed), further purification may then be desirable. This can be accomplished very easily by centrifuging the resuspended chloroplasts through a cushion of Percoll, which is non-toxic and which has a very low osmotic potential.

(i) Put 5 ml of 40% (v/v) Percoll (in resuspension medium, a lower concentration of Percoll may be required) into the centrifuge tube, carefully overlay it with the chloroplast suspension and centrifuge for 1 min.

(ii) Discard the supernatant and resuspend the pellet in fresh medium.

The chloroplasts purified in this way should be almost wholly intact (19,20). More rigorous purification, yielding extremely stable (up to 3 days) and highly active chloroplasts entirely free from contaminating organelles, may be accomplished by centrifugation in a Percoll gradient (21,22).

The mechanical procedures and media developed for the isolation of spinach chloroplasts have been satisfactorily modified to suit other species such as peas (1,2,4,23), wheat (24) and maize (19). A suitable grinding and washing medium for peas comprises: 0.33 M glucose, 50 mM Na_2HPO_4, 50 mM K_2HPO_4, 5 mM $MgCl_2$, 0.1% (w/v) NaCl, 0.2% (w/v) sodium iso-ascorbate, 0.1% (w/v) BSA, adjusted to pH 6.5 with HCl, with resuspension in the same medium as for spinach. Media low in cations have also found favour due to an improvement in chloroplast intactness which results from an enhanced separation of intact and broken chloroplasts. A combination of MES and Tris buffers has been used to achieve a low ionic strength (25), but chloroplast stability is sacrificed (11). However, if the intactness is already high, little improvement can be detected. A new low cation grinding and washing medium for peas has recently been described (1) in which the buffer concentration is merely reduced to a minimum, comprising 343 mM sorbitol, 0.4 mM KCl, 0.04 mM EDTA, 2 mM Hepes (pH 7.8), in a procedure essentially the same as that described in *Table 2*. The resuspension medium is the same as that for spinach [*Table 2*, (11)].

3.1.2 *Preparation of Chloroplasts from Protoplasts*

Protoplasts from C_3 and C_4 plants are usually between 30 and 40 μm in diameter.

Table 3. Preparation of Chloroplasts from Wheat Protoplasts.

1. Pellet protoplasts which contain the equivalent of 200–400 μg chlorophyll by centrifugation at 100 *g* for 2 min. 5-ml plastic tubes are ideal for this purpose.
2. Discard the supernatant and resuspend the pellet in 0.5 ml ice-cold break medium [0.4 M sorbitol, 10 mM EDTA, 25 mM Tricine-KOH (pH 8.0)].
3. Take the protoplasts up through a 20 μm nylon mesh attached to the end of a plastic syringe and expel them. Repeat this procedure twice more.
4. Centrifuge at 250 *g* for 45 sec, discard the supernatant and resuspend the pellet in the above medium.

Chloroplasts are easily prepared from protoplasts by passing them through a 20 μm mesh (*Table 3*). First, pellet the protoplasts (not more than the equivalent of about 400 μg chlorophyll) by centrifugation (100 *g* for 2 min). If appreciable amounts of medium II have been carried over from the purification step, then addition of medium I may be required, since some of the protoplasts may remain floating. After centrifugation, discard the supernatant and resuspend the protoplasts by gentle shaking in the break medium. A simple device can be made by cutting the end from a 1 ml plastic syringe and attaching a small square of 20 μm mesh using a plastic ring (1–2 mm wide) cut from a plastic automatic pipette tip. Wash out the syringe with alcohol to remove the lubricant. The nylon mesh (Nybolt P-20) can be obtained from Schweizerische Seidengazefabrik, Zurich, Switzerland.

After centrifugation discard the supernatant and resuspend the pellet by gentle agitation in fresh medium. Sometimes aggregated material may overlie the pellet; gently dislodge this with resuspension medium and discard before resuspension of the remainder of the pellet.

3.2 Assay for Chloroplast Intactness

The simplest assay for intactness is a comparison of the rate of reduction of ferricyanide (which does not permeate the chloroplast envelope) in two samples of the chloroplast preparation, only one of which has been subjected to osmotic shock (13). The uncoupled rate of ferricyanide reduction is followed in an O_2 electrode (Chapter 5).

(i) For the intact preparation, add 0.1 ml chloroplast suspension (not >100 μg chlorophyll) to 1.9 ml of assay medium [0.33 M sorbitol, 2 mM EDTA, 1 mM $MgCl_2$, 1 mM $MnCl_2$, 50 mM Hepes-KOH (pH 7.6)], together with 3 mM $K_3Fe(CN)_6$ and 10 mM D,L-glyceraldehyde (the latter inhibits CO_2 fixation, and prevents possible interference).

(ii) Add 5 mM NH_4Cl (an uncoupler) a minute or two after illumination.

(iii) For the osmotically-shocked preparation, add 0.1 ml of chloroplast suspension to 0.9 ml of 10 mM $MgCl_2$ and stir for 1 min.

(iv) Add 1 ml of double strength assay medium, together with the same additions.

(v) The percentage intactness is given by the ratio of the rates of uncoupled electron transport in the two preparations. For example, if the rates were 250 and 30 μmol (mg chl)$^{-1}$ h^{-1} chlorophyll in the shocked and intact preparations, respectively, then the chloroplast intactness would be 100–[30 × 100/250] = 88%. The assay probably overestimates intactness due to rupture and resealing of chloroplasts (13).

3.3 Assay for CO_2-dependent O_2 Evolution by Chloroplasts

Good chloroplasts can be expected to fix CO_2 at rates in excess of 100 μmol (mg chl)$^{-1}$ h^{-1} when illuminated in a suitable medium at 20°C. Since the isolated chloroplast is a phosphate-importing, triose phosphate-exporting organelle, Pi must also be included. With mechanically-prepared spinach chloroplasts the standard resuspension medium (26) (*Table 2*) is best supplemented with 10 mM $NaHCO_3$, 0.5 mM Pi and catalase (1000 U ml^{-1}). In practice the Pi optimum is difficult to achieve and the addition of 5 mM pyrophosphate may be beneficial, since in the presence of Mg^{2+} and contaminating pyrophosphatase, PPi acts as an optimal source of Pi (27), although it is inhibitory to wheat and pea chloroplasts in the absence of added adenylates (2,3,23). Some manipulation of Pi concentration, pH etc., may be required to obtain optimum rates. With chloroplasts prepared from such species as wheat, sunflower and maize from protoplasts, the use of high concentrations of EDTA in the medium is advisable (14,28) and a more alkaline pH may be required [e.g. 0.4 M sorbitol, 10 mM EDTA, 50 mM Tricine (pH 8.0) supplemented with 10 mM $NaHCO_3$, 1000 U ml^{-1} catalase and 0.2 mM Pi for wheat chloroplasts].

If chloroplasts fail to fix CO_2 at high rates, then their inability to evolve O_2 in the presence of 3-phosphoglycerate is indicative of very poor quality chloroplasts.

4. ISOLATION OF SUB-CHLOROPLAST FRACTIONS

4.1 Preparation of Chloroplasts and Thylakoid Membranes

The isolation of chloroplasts as a first stage in the preparation of sub-chloroplast fractions need not be performed with the great care that has been described in Section 3. Nevertheless, there are advantages to be gained from starting from a batch of chloroplasts with 70–80% intactness, followed by osmotic lysis. The isolation of intact chloroplasts can be accomplished on a large scale by using mechanical grinding of plant material in a medium of high ionic strength.

All steps and centrifugations should be carried out at 4°C unless otherwise stated.

(i) Prepare a grinding solution that contains 0.3 M sucrose, 50 mM Tris-HCl buffer (pH 7.8), 20 mM NaCl and 5 mM $MgCl_2$.

(ii) Blend 50 g of freshly harvested spinach or pea leaves in 150 ml of the grinding solution at 4°C for 20 sec.

(iii) Filter the slurry through filtering silk (Miracloth, Calbiochem) and centrifuge the resulting slurry at 3000 *g* for 60 sec.

(iv) Discard the supernatant solution and resuspend the precipitate in a small volume of grinding solution. This fraction contains ~70% intact chloroplasts based on measurements of ferricyanide reduction (13).

This procedure provides a rapid means for the preparation of intact chloroplasts essentially free from contamination of other cellular organelles, such as mitochondria. Thylakoid membranes, free of stromal contamination, may be isolated by osmotically disrupting the outer chloroplast membrane and re-isolating the thylakoid membranes by centrifugation.

To eliminate stromal proteins, such as ribulose bisphosphate carboxylase, the intact chloroplasts are suspended in a high ionic strength medium buffered at approximately pH 8.0 (10 mM Tricine-KOH buffer, pH 8.0 + 150 mM NaCl) at 4°C. Use approximately 100 ml of this solution for about 20 mg of chlorophyll. The suspension is gently mixed and the membranes re-isolated by centrifugation at 35 000 *g* for 5 min. The pellet, which contains thylakoid membranes, is resuspended in a small volume of the original grinding solution.

Thylakoid membranes prepared by this procedure should be used immediately as photochemical activity decreases as a function of time. These membranes are capable of high rates of oxygen evolution coupled to the reduction of electron acceptors such as NADP or ferricyanide (Chapter 5). These membranes also show excellent coupling in that electron transport is coupled to the synthesis of ATP (Chapter 6). Membranes prepared by this procedure are routinely used for the preparation of photosynthetic membrane complexes, as described in greater detail below.

It should be realised that there is no single 'correct' method for the isolation of thylakoid membranes: the technique used will depend to a very large extent on the use to which the membranes will be put. Thus, in addition to the method described above, methods for the isolation of thylakoid membranes are also given in Chapters 3 and 6.

4.2 **Photosystem I Sub-chloroplast Fraction**

The photosystem I (PSI) complex is prepared from spinach thylakoid membranes suspended in a low cation medium by solubilisation with low concentrations of the detergent Triton X-100 (29). The complex is isolated by centrifugation on sucrose gradients.

4.2.1 *Preparation of Spinach Thylakoid Membranes*

Intact chloroplasts are prepared from spinach leaves as described in Section 4.1. The chloroplast pellet should contain 20–30 mg of chlorophyll.

4.2.2 *Triton X-100 Solubilisation*

(i) Resuspend the pellet in 50 mM sorbitol and 5 mM EDTA (pH 7.8) (~150 ml) and re-isolate the thylakoid membranes by centrifugation at 10 000 *g* for 10 min.

(ii) Resuspend the pellet in distilled water and add Triton X-100 to give a final concentration of approximately 0.8%.

The Triton is added from a 20% (w/v) solution. This solution is prepared by weighing 20 g Triton and bringing the final volume to 100 ml with water. It is essential to stir this suspension for a sufficient length of time for all the Triton to dissolve. The final chlorophyll concentration should be 0.8 mg ml^{-1}. The exact amount of Triton to be added is critical and should produce a 90% solubilisation of the membranes based on the amount of chlorophyll recovered in a 42 000 *g* pellet. This amount of Triton should be determined on a small aliquot of the membrane suspension using varying amounts of Triton from a stock 20% (w/v) solution. Solubilisation is estimated by the amount of chlorophyll recovered after a 30 min centrifugation at 42 000 *g* and, as shown in *Table 4*, over 90% solubilisation is usually achieved with 0.8% Triton, but this can vary depending on the exact chlorophyll and Triton concentrations during the incubation. Seasonal variations also exist in relation to this important factor, so the exact amount of Triton to be added should be determined periodically. Excess Triton is to be avoid-

Table 4. Solubilisation of Chloroplast Membranes with Triton X-100[a].

Triton concentration used for solubilisation %	*Chlorophyll recovered in pellet* % *of total*
0.4	30
0.6	24
0.7	14
0.8	6
0.9	3

[a]Membranes were isolated as described in the text and were resuspended in H_2O in the presence of the indicated amount of Triton at a chlorophyll concentration of 0.8 mg ml^{-1}. After stirring for 30 min at 20°C, the suspension was centrifuged at 42 000 *g* for 30 min and the amount of chlorophyll in the pellet and supernatant solution determined [From ref. (29).]

ed since this releases free chlorophyll from PSI, as indicated by fluorescence in the 670 – 675 nm spectral region (Chapter 4).

4.2.3 *Sucrose Gradient Centrifugation*

(i) After the addition of the Triton X-100, stir the chloroplast suspension slowly at 20°C for 30 min and then centrifuge at 42 000 *g* for 30 min.
(ii) Discard the pellet and fractionate the green supernatant solution by sucrose gradient centrifugation. The principles of gradient centrifugation, as well as practical details, have been covered in a previous volume in this series (30).
(iii) Load 8 ml of the supernatant solution on 25 ml of a 0.1 – 1.0 M sucrose gradient (prepared in the presence of 0.05% Triton X-100) which was layered on a cushion containing 5 ml of 2 M sucrose in 0.05% Triton X-100.
(iv) Centrifuge the sucrose gradient at 100 000 *g* for 15 h in a Spinco SW-27 or SW-25 rotor.

The PSI complex is localised as a dark green band at the interface between the 1 M and 2 M sucrose solutions and a large amount of chlorophyll is present in the upper portion of the gradient. The upper portion of the gradient is discarded and the PSI complex is collected. Store the PSI complex at –20°C in small aliquots.

The PSI complex has a chlorophyll *a/b* ratio of approximately 5.5. The P700 content is 1 per approximately 220 chlorophylls based on oxidised-minus-reduced difference spectroscopy of P700 measured at 435 – 444 nm, with a millimolar extinction coefficient of 44 (Chapter 4). The preparation also shows photosystem I activity, such as electron transport from reduced dichlorophenol indophenol to methyl viologen (Chapter 5), and oxygen uptake rates of the order of 200 μmol (mg chl)$^{-1}$ h^{-1} are typical.

4.3 **Photosystem II Sub-chloroplast Fraction**

The photosystem II (PSII) complex is prepared from spinach thylakoid membranes by solubilisation with a high concentration of Triton X-100 in the presence of cations and the complex isolated by differential centrifugation (31).

4.3.1 *Preparation of Spinach Thylakoid Membranes*

Isolate intact chloroplasts from spinach leaves by the method given in Section 4.1, but with the grinding medium modified by the addition of 0.2% BSA. The intact chloroplast pellet should contain 20 – 30 mg of chlorophyll. Resuspend the pellet in about 150 ml of a solution containing 20 mM Hepes buffer (pH 7.5), 5 mM $MgCl_2$, 150 mM NaCl and 0.2% BSA. Re-isolate the membranes by centrifugation at 35 000 *g*.

4.3.2 *Triton X-100 Treatment*

(i) Resuspend the thylakoid pellet in a solution containing 20 mM MES buffer (pH 6.0), 20 mM NaCl, 5 mM $MgCl_2$ and 0.2% BSA.
(ii) Add Triton X-100 from a stock 20% (w/v) solution to give 25 mg Triton (mg chl)$^{-1}$ and a final chlorophyll concentration of 2 mg ml^{-1}.
(iii) Incubate this suspension at 4°C for 30 min with occasional mixing. After this period, centrifuge the suspension at 40 000 *g* for 30 min.
(iv) Immediately resuspend the pellet in the same solution as in the previous step and add Triton X-100 to give 5 mg Triton (mg chl)$^{-1}$ and a chlorophyll concentration of 2 mg ml^{-1}.
(v) Mix this suspension and immediately centrifuge at 40 000 *g* for 30 min.
(vi) Resuspend the pellet in the same buffer as before that includes 0.3 M sucrose and store aliquots of this suspension at −20°C.

The chlorophyll *a*/*b* ratio of the PSII fraction should be about 1.8, and the final yield of chlorophyll in the PSII fraction is approximately 50% of the starting amount of chlorophyll. The PSII fraction catalyses oxygen evolution with a suitable electron acceptor, such as dimethylbenzoquinone (Chapter 5). Rates of oxygen evolution are usually in the range 200 – 400 μmol (mg chl)$^{-1}$ h^{-1}.

5. ISOLATION OF FUNCTIONAL COMPLEXES

5.1 The Cytochrome b_6-*f* Complex

The cytochrome b_6-*f* complex is prepared from thylakoid membranes that have been extensively washed to remove peripheral proteins and coupling factor. The complex is solubilised with the detergents octyl glucoside and sodium cholate and subsequently purified by ammonium sulphate fractionation and centrifugation on a sucrose gradient (32,33).

5.1.1 *Preparation of Spinach Thylakoid Membranes*

(i) Isolate intact chloroplasts from spinach leaves as described in Section 4.1. The intact chloroplast pellet should contain 20 – 30 mg of chlorophyll.
(ii) Resuspend the pellet in 150 ml of 0.15 M NaCl to disrupt the outer chloroplast membrane and centrifuge this solution for 10 min at 13 000 *g* to pellet the thylakoid membranes.
(iii) Resuspend the pellet in grinding solution that contains 2 M NaBr to a final chlorophyll concentration of 1 mg ml^{-1}. This treatment removes peripheral proteins and the chloroplast coupling factor.
(iv) Incubate for 30 min at 4°C and, after this time, add an equal volume of cold distilled water.
(v) Centrifuge at 13 000 *g* for 20 min. Treat the pellet in the same manner to remove

any residual coupling factor. After the second sodium bromide wash, resuspend the pellet in 150 ml of the grinding solution and centrifuge at 13 000 *g* for 10 min to obtain a pellet containing well-washed thylakoid membranes.

At this stage, the chloroplast pellet can be stored overnight at 4°C prior to the addition of the detergents, but it is preferable to continue the purification and carry it to completion in 1 day.

5.1.2 *Extraction of the Cytochrome Complex*

(i) Resuspend the pellet containing the thylakoid membranes in grinding solution to give a final chlorophyll concentration of 3 mg ml^{-1}.

(ii) Add an equal volume of grinding solution that contains 60 mM octyl glucoside, 1% sodium cholate and 0.8 M ammonium sulphate. This will yield final concentrations of 1.5 mg ml^{-1} of chlorophyll, 30 mM octyl glucoside, 0.5% cholate and approximately 5% ammonium sulphate.

(iii) Incubate this suspension at 4°C for 30 min and then centrifuge for 1 h at 360 000 *g*.

The supernatant solution contains the solubilised cytochrome complex and, when assayed for cytochrome *f* by ascorbate-reduced minus ferricyanide-oxidised difference spectrophotometry (using ϵ_{mM} = 20 at 554 – 540 nm), contains approximately 1.5 nmol cytochrome *f* per mg chlorophyll in the starting material (Chapter 4). Approximately 10% of the starting amount of chlorophyll will also be solubilised by this treatment.

5.1.3 *Ammonium Sulphate Fractionation*

(i) Prepare a saturated ammonium sulphate solution by dissolving solid ammonium sulphate in distilled water at 25°C and adjust the pH of this solution to 7.5.

(ii) Add ammonium sulphate from the saturated solution to the supernatant solution containing the solubilised cytochrome complex to give 45% saturation.

(iii) Stir the suspension at 4°C for 10 min and centrifuge for 10 min at 13 000 *g*.

(iv) Discard the pellet and add ammonium sulphate to the supernatant solution to give 55% saturation.

(v) Stir for 10 min at 4°C, then centrifuge the suspension at 13 000 *g* for 10 min.

(vi) Discard the supernatant solution and redissolve the pellet in a minimal amount of a solution containing 30 mM Tris-succinate buffer (pH 6.5) and 0.5% sodium cholate.

Tris-succinate buffer is prepared by dissolving solid Tris base in distilled water and adjusting the pH to 6.5 with succinic acid. A 300 mM stock solution is usually prepared and diluted to 30 mM prior to use. Only 1 – 2 ml of buffer should be required.

(vii) The cytochrome *f* concentration of this solution, estimated by reduced-minus-oxidised difference spectrophotometry, should be 20 – 40 μM. Transfer the solution to a small dialysis bag, place in a flask containing 30 mM Tris-succinate buffer plus cholate, and gently stir the solution at 4°C. It is critical not to dialyse the solution for longer than 45 min or protein aggregation will occur and the cytochrome complex will precipitate from solution.

5.1.4 *Sucrose Gradient Centrifugation*

(i) Apply aliquots of the dialysed solution to a 7–30% (w/v) linear sucrose gradient (30) prepared in the presence of 30 mM Tris-succinate buffer (pH 6.5) and 0.5% sodium cholate. Use gradients of approximately 10 ml volume and apply 1 ml of sample per tube.

(ii) Centrifuge the gradients at 300 000 *g* for 16 h in a Spinco SW-41 rotor. At the end of this centrifugation, the cytochrome complex will be visualised as a brown band in the middle of the gradient.

(iii) Collect the band with a syringe and pool the fractions from individual tubes. The cytochrome *f* content, estimated from reduced-minus-oxidised difference spectra, is usually 10–20 μM, and the overall yield based on cytochrome *f* is 40–50%.

(iv) Store the complex in small aliquots by freezing samples rapidly to 77 K. Storage is best done at 77 K but it may be more convenient to store at −20°C.

The final preparation should contain cytochrome b_6 and cytochrome *f* in a ratio of 2:1. The former cytochrome is estimated from dithionite-minus-ascorbate reduced difference spectra, using an extinction coefficient of 20 mM cm^{-1} at 563–575 nm. Polyacrylamide gel electrophoresis on SDS gels (SDS-PAGE; Chapter 3, ref. 34) should show the presence of only four polypeptide subunits: 33 kd (cytochrome *f*), 23 kd (cytochrome b_6), 20 kd (the Rieske Fe-S centre) and 17 kd.

This procedure has been carried out on chloroplast preparations as small as 5 mg and as large as 1000 mg although in the latter case it is difficult to carry out the entire procedure in a single day and several centrifugation steps are required for the final sucrose gradient centrifugation.

5.2 **PSI Reaction Centre Core Complex**

The resolved PSI complex, prepared as described in Section 4.2, is used for the preparation of the PSI reaction centre core complex. The procedure involves treatment of the native PSI complex with the detergents dodecyl-β-D maltoside and Zwittergent 3-16 (Calbiochem), and purification of the complex by sucrose gradient centrifugation (35).

(i) Dialyse the PSI preparation at 4°C overnight against 50 mM sorbitol to remove Triton X-100 and sucrose from the final sucrose gradient centrifugation step.

(ii) Re-isolate the particles by centrifugation at 40 000 *g* for 10 min.

(iii) Resuspend the particles to a final chlorophyll concentration of 0.5 mg ml^{-1} in the presence of 2 mg ml^{-1} of Zwittergent 3-16 and 1.5 mg ml^{-1} of dodecyl-β-D-maltoside.

(iv) Stir at 4°C for 60 min.

(v) The reaction centre core complex is isolated by sucrose gradient centrifugation. Apply 3 ml of detergent-treated PSI to a 13 ml gradient (0.1–1.0 M sucrose containing 20 mM Tricine-NaOH, pH 7.8, plus 1% dodecyl-β-D-maltoside, overlaid on a 2 ml cushion of 2 M sucrose), and centrifuge the gradient in a Beckman SW-28 rotor for 15 h at 100 000 *g*.

Alternatively, gradients can be prepared in the fixed angle tubes of a Ti60 rotor, and these gradients centrifuged at 360 000 *g* for 3 h. The lower green band in these gradients contains the PSI reaction centre core complex while the upper green band contains a light-harvesting chlorophyll *a/b* complex associated with PSI (Chapter 3).

The PSI reaction centre core complex prepared by the above method contains a ratio of chlorophyll to P700 of 100:1 based on photoinduced redox changes of P700. The chlorophyll *a/b* ratio of this complex is in excess of nine while the same ratio for the native PSI complex is approximately 5.5, due to the presence of the chlorophyll *a/b* complex in the native complex. The polypeptide composition of the PSI core complex, determined by SDS-PAGE analysis (Chapter 3; ref. 34) in 10–20% gradient gels by the method of Chua (36), is relatively simple: ~65-, ~19-, ~14- and ~10-kd polypeptides are present. E.p.r. measurements at cryogenic temperatures (Chapter 4) have demonstrated the presence of iron-sulphur centres A and B of the PSI primary electron acceptor complex.

5.3 **PSII Reaction Centre Core Complex**

The oxygen-evolving PSII complex, described in Section 4.3, is used as the starting material for the PSII reaction centre core complex (37,38).

(i) Suspend the pellet from the second Triton treatment of the PSII preparation in 1 M Tris-HCl (pH 9.0) using 80 ml of buffer for 20 min at 4°C and re-isolate the PSII fragments by centrifugation at 40 000 *g* for 10 min. Repeat this treatment one additional time. This step removes extraneous proteins as well as three PSII polypeptides of molecular weights of approximately 16, 23 and 33 kd.
(ii) Resuspend the pellet from the buffer treatment in 0.05 M MES buffer (pH 6.0) at a chlorophyll concentration of 1 mg ml^{-1} and add dodecyl-β-D-maltoside to a detergent/chlorophyll ratio (w/v) of 10:1.
(iii) Incubate at 4°C for 20 min and centrifuge at 1000 *g* for 1 min to remove starch.
(iv) Layer the supernatant on a 10–30% sucrose gradient prepared in 50 mM MES buffer (pH 6.5) + 0.05% Triton. Use approximately 11 ml sucrose gradient volumes and apply 1 ml of sample to each tube in a Spinco SW-41 rotor.
(v) Centrifuge for 24 h at 300 000 *g*. After this time, two green bands should be present: the broad upper band contains the PSII light-harvesting chlorophyll *a/b* complex while the smaller lower band contains the PSII reaction centre core complex. Occasionally a pale lower band, which contains residual PSI, is present below the PSII core.
(vi) Carefully remove the PSII core complex, trying to separate it from the chlorophyll *a/b* protein. Store in small aliquots at –20°C.

The PSII reaction centre core complex no longer evolves oxygen but is active in catalysing electron transfer from diphenylcarbazide to dichlorophenolindophenol (Chapter 5). SDS-PAGE analysis (Chapter 3; ref. 34) by the method of Chua (36) indicates the presence of four major polypeptides of 47, ~44, ~33 and 10 kd. As an indication of purity, no polypeptides in the 25–29 kd range (PSII chlorophyll *a/b* protein) should be present in the preparation.

5.4 **The Chloroplast Coupling Factor**

5.4.1 *The ATP Synthetase Complex (CF_0-CF_1)*

The ATP synthetase complex is composed of two portions: a hydrophobic portion (CF_0)

and a hydrophilic portion (CF_1) (Chapter 6). The intact complex containing both parts can be isolated from thylakoid membranes by detergent solubilisation, followed by fractionation with ammonium sulphate and sucrose gradient centrifugation (39).

(i) *Preparation of spinach thylakoid membranes.*

(a) Isolate intact chloroplasts from spinach leaves as described in Section 4.1. The chloroplast pellet should contain 20–30 mg of chlorophyll.

(b) Resuspend the pellet in 150 ml of a solution containing Tricine-NaOH buffer (pH 8.0) and 0.15 M NaCl to disrupt the chloroplast outer membrane and remove ribulose bisphosphate carboxylase; re-isolate the thylakoid membranes by centrifugation at 35 000 *g* for 5 min.

(c) Resuspend the pellet in the original grinding solution to give a final chlorophyll concentration of 4 mg ml^{-1}.

(ii) *Solubilisation of the* CF_0-CF_1 *complex.*

(a) Add 1 ml of 0.5 M dithiothreitol (DTT) per 10 ml of chloroplast suspension and leave for 15 min at 4°C.

(b) Add an equal volume of grinding solution that contains 60 mM octylglucoside, 1% sodium cholate, 100 mM DTT and 10% ammonium sulphate.

(c) Incubate this suspension at 4°C for 15 min and then centrifuge at 230 000 *g* for 60 min. The supernatant, which should be straw-coloured, contains the solubilised CF_0-CF_1 complex.

(iii) *Purification of the* CF_0-CF_1 *complex.*

(a) Prepare a saturated ammonium sulphate solution by dissolving ammonium sulphate in distilled water at 25°C and adjust the pH to 8.0.

(b) Add ammonium sulphate to the supernatant solution from the high-speed centrifugation to give a final concentration of 37.5% (0.44 ml saturated ammonium sulphate per ml).

(c) Leave for 10 min at 4°C and centrifuge for 10 min at 13 000 *g*.

(d) Add ammonium sulphate to the supernatant solution to give 45% saturation (an additional 0.136 ml of saturated ammonium sulphate solution per ml).

(e) Allow this mixture to stand for 20 min at 4°C and isolate the precipitate by centrifugation at 13 000 *g* for 10 min.

(f) Carefully remove all the supernatant solution and dry the walls of the centrifuge tubes.

(g) Dissolve the pellet in a minimal volume of a solution that contains 0.2 M sucrose, 20 mM Tricine-NaOH buffer (pH 8.0) and 3 mM $MgCl_2$.

This ammonium sulphate fraction can be stored at 77 K for use as a crude ATP-synthetase or can be further purified by sucrose gradient centrifugation.

(h) Dilute a sample of the ammonium sulphate fraction containing 3–5 mg of protein to 0.5 ml with a solution containing 20 mM Tris-succinate buffer (pH 6.5), 0.5 mM sodium-EDTA, 0.1 mM ATP and 0.2% Triton X-100.

(i) Layer onto four sucrose gradient tubes for a Spinco SW-60Ti rotor.

Two possible gradients can be used: a linear 7 – 30% sucrose gradient made in the presence of 0.1% sonicated soybean phospholipids in the above buffer or a linear 7 – 40% sucrose gradient with 0.4% sodium cholate instead of Triton X-100 and 0.1% sonicated phospholipids in the above medium.

(j) Centrifuge for 5 h at 150 000 *g*. Collect 0.25 ml fractions and store at 77 K.
The peak of the ATP synthetase activity should appear in the Triton gradient around 40% down the tube (fraction 7 – 9 from the top) and in the cholate gradients around 70% down the tube (fraction 11 – 15 from the top).

The purified ATP synthetase contains eight polypeptides on SDS-PAGE analysis (34). Polypeptides of molecular weights 59, 55, 37.5, 17.5 and 13 kd are associated with CF_1 while polypeptides of molecular weights 15.5, 13 and 7.5 kd are associated with CF_0. The latter low molecular weight polypeptide has been shown to be the 1,3-dicyclohexyl carbodiimide (DCCD) binding subunit of the ATP synthetase complex.

5.4.2 *The Coupling Factor* (CF_1)

Various methods are available for the preparation of the water-soluble portion of the ATP synthetase complex, known as CF_1. A relatively straightforward procedure involves extensive washing of thylakoid membranes to remove extraneous soluble proteins, followed by release of the coupling factor with hypertonic sucrose solution (40).

(i) *Preparation of spinach thylakoid membranes.*

(a) Prepare intact spinach chloroplasts as described in Section 4.1.
(b) Resuspend the intact chloroplast pellet in 150 ml of 10 mM sodium pyrophosphate (pH 7.4) and re-isolate the membranes by centrifugation at 35 000 *g* for 10 min.
(c) Repeat this treatment three additional times to remove all ribulose bisphosphate carboxylase and other stromal proteins. These washed membranes are then used for extraction of CF_1.

(ii) CF_1 *purification.*

(a) Resuspend the pyrophosphate-washed membranes in 0.3 M sucrose + 2 mM Tricine-NaOH buffer (pH 7.8).
(b) Re-isolate the membranes by centrifugation at 35 000 *g* for 10 min and save the supernatant solution, which contains CF_1.

Approximately half of the membrane-bound CF_1 has been found to be released in the supernatant by this treatment and the resulting solubilised CF_1 prepared by this procedure is essentially pure when analysed by SDS-PAGE (34).

CF_1 can be assayed by rebinding the enzyme to depleted membranes and measuring light-dependent ATP formation (Chapter 6). It is, however, more convenient to measure the latent Ca^{2+}-dependent ATPase activity of the enzyme. This activity is present after heat treatment, after trypsin digestion, or after prolonged incubation with DTT. The activity is assayed by determining the amount of inorganic phosphate liberated from ATP (41; Chapter 6). The pure protein catalyses the hydrolysis of ~30 μmol ATP (mg protein)$^{-1}$ min^{-1} after trypsin activation.

6. REFERENCES

1. Cerovic,Z.G. and Plesnicar,M. (1984) *Biochem. J.*, **223**, 543.
2. Robinson,S.P. and Wiskich,J.T. (1976) *Plant Physiol.*, **58**, 156.
3. Robinson,S.P. and Wiskich,J.T. (1977) *Plant Physiol.*, **59**, 422.
4. Walker,D.A. (1980) in *Methods in Enzymology*, Vol. **69**, San Pietro,A. (ed.), Academic Press, New York p. 94.
5. Edwards,G.E., Robinson,S.P., Tyler,N.J.C. and Walker,D.A. (1978) *Plant Physiol.*, **62**, 313.
6. Leegood,R.C. and Walker,D.A. (1983) in *Isolation of Membranes and Organelles from Plant Cells*, Hall,J.L. and Moore,A.L. (eds.), Academic Press, New York, p. 185.
7. Kanai,R. and Edwards,G.E. (1973) *Plant Physiol.*, **52**, 484.
8. Chapman,K.S.R., Berry,J.A. and Hatch,M.D. (1980) *Arch. Biochem. Biophys.*, **202**, 330.
9. Edwards,G.E., Lilley,R.McC., Craig,S. and Hatch,M.D. (1979) *Plant Physiol.*, **63**, 821.
10. Moore,B.D., Ku,M.S.B. and Edwards,G.E. (1984) *Plant Sci. Lett.*, **35**, 127.
11. Jensen,R.G. (1979) in *Encyclopedia of Plant Physiology*, Vol. **2**, Gibbs,M. and Latzko,E. (eds.), Springer Verlag, Berlin, p. 31.
12. Nagata,T. and Ishii,S. (1979) *Can. J. Bot.*, **57**, 1820.
13. Lilley,R.McC., Fitzgerald,M.P., Rienits,K.G. and Walker,D.A. (1975) *New Phytol.*, **75**, 1.
14. Day,D.A., Jenkins,C.L.D. and Hatch,M.D. (1980) *Aust. J. Plant Physiol.*, **8**, 21.
15. Nishimura,M., Hara-Nishimura,I. and Robinson,S.P. (1984) *Plant Sci. Lett.*, **37**, 171.
16. Edwards,G.E., Huber,S.C. and Gutierrez,M. (1976) in *Microbial and Plant Protoplasts*, Pederby,J.F., Rose,A.H., Rogers,J.H. and Cocking,E.C. (eds.), Academic Press, New York, p. 299.
17. Huber,S.C. and Edwards,G.E. (1975) *Physiol. Plant.*, **35**, 203.
18. Lilley,R.McC. and Walker,D.A. (1979) in *Encyclopedia of Plant Physiology*, Vol. **2**, Gibbs,M. and Latzko,E. (eds.), Springer Verlag, Berlin, p. 41.
19. Jenkins,C.L.D. and Russ,V.J. (1984) *Plant Sci. Lett.*, **35**, 19.
20. Robinson,S.P. (1982) *Plant Physiol.*, **70**, 1032.
21. Mills,W.R. and Joy,K.W. (1980) *Planta*, **148**, 75.
22. Mourioux,W.R. and Douce,R. (1981) *Plant Physiol.*, **67**, 470.
23. Stankovic,Z.S. and Walker,D.A. (1977) *Plant Physiol.*, **59**, 428.
24. Leegood,R.C. and Walker,D.A. (1979) *Plant Physiol.*, **63**, 1212.
25. Nakatani,H.Y. and Barber,J. (1977) *Biochim. Biophys. Acta*, **461**, 510.
26. Jensen,R.G. and Bassham,J.A. (1966) *Proc. Natl. Acad. Sci. USA*, **56**, 1095.
27. Schwenn,J.D., Lilley,R.McC. and Walker,D.A. (1973) *Biochim. Biophys. Acta*, **325**, 586.
28. Edwards,G.E., Robinson,S.P., Tyler,N.J.C and Walker,D.A. (1978) *Arch. Biochem. Biophys.*, **190**, 421.
29. Mullet,J.E., Burke,J.J. and Arntzen,C.J. (1980) *Plant Physiol.*, **65**, 814.
30. Rickwood,D., ed. (1984) *Centrifugation – A Practical Approach* (2nd edn.), published by IRL Press, Oxford and Washington, D.C.
31. Berthold,D.A., Babcock,G.T. and Yocum,C.F. (1981) *FEBS Lett.*, **134**, 231.
32. Hurt,E. and Hauska,G. (1981) *Eur. J. Biochem.*, **117**, 591.
33. Hurt,E. and Hauska,G. (1982) *J. Bioenerg. Biomembr.*, **14**, 119.
34. Hames,B.D. (1981) in *Gel Electrophoresis of Proteins – A Practical Approach*, Hames,B.D. and Rickwood,D. (eds.), IRL Press, Oxford and Washington, D.C., p. 1.
35. Haworth,P., Watson,J.L. and Arntzen,C.J. (1983) *Biochim. Biophys. Acta*, **724**, 151.
36. Chua,N.-H. (1980) in *Methods in Enzymology*, Vol. **69**, San Pietro,A. (ed.), Academic Press, New York, p. 434.
37. Westhoff,P., Alt,J. and Hermann,R.G. (1983) *EMBO J.*, **2**, 2229.
38. Bricker,T.M., Pakrasi,H.B. and Sherman,L.A. (1984) *Plant Physiol.*, **75**, 1015.
39. Pick,U. and Racker,E. (1979) *J. Biol. Chem.*, **254**, 2793.
40. Strotman,H., Hesse,H. and Edelmann,K. (1973) *Biochim. Biophys. Acta*, **314**, 202.
41. Lien,S. and Racker,E. (1971) in *Methods in Enzymology*, Vol. **23A**, San Pietro,A. (ed.), Academic Press, New York, p. 547.

CHAPTER 3

Electrophoretic Analysis of Photosynthetic Pigment-protein Complexes

J. MARKWELL

1. INTRODUCTION

In mature leaves, the photosynthetic pigments, carotenoids and chlorophylls *a* and *b*, are localised within the network of membranes within the chloroplast termed 'thylakoid' membranes. Since these pigments are hydrophobic in nature, it was long assumed that they were dissolved into the lipid phase of the thylakoid membrane (cf. 1). However, work in the past 20 years (2 – 4) has demonstrated that these molecules exist in the form of pigment-protein complexes (PPCs). It is now known that practically all of the chlorophyll in the thylakoid membrane is normally bound to protein rather than existing free in the lipid phase (5) and that all of the biophysical functions of light-harvesting, energy transfer and photochemical energy conversion carried out by the photosynthetic pigments are performed by PPCs.

These photosynthetic PPCs are the major components of the thylakoid membrane and are generally hydrophobic in nature. The bulk of the PPCs functions solely as antenna components within the photosynthetic apparatus, harvesting light energy and transferring it to the photochemical reaction centres which contain only a few percent of the pigment molecules. Since most of the PPCs contain the same pigments in similar proportions, their absorption spectra are frequently similar. Thus there are usually no specific or direct criteria for identifying most PPCs when purified or to judge progress during a purification. For these reasons, studies on the characteristics and functions of the PPCs which make up the photosynthetic apparatus have not progressed as rapidly as studies on soluble pigment-protein complexes such as haemoglobin or the cytochromes.

The method that has had the most success in studies on the photosynthetic PPCs is polyacrylamide gel electrophoresis following solubilisation of the thylakoid membrane by surfactant treatment. This chapter will not attempt to provide a detailed or theoretical treatment of polyacrylamide gel electrophoresis, which has already been thoroughly discussed in this series (6), or of the use of surfactants in membrane studies. Rather, it will attempt to present a practical approach that will permit a researcher with little previous experience in these techniques to fractionate higher plant photosynthetic PPCs and to understand the utility and limitations of this technique.

2. LIMITATIONS OF THE TECHNIQUE

Successful electrophoretic fractionation requires that a fine balance be achieved. For example, treatment with insufficient surfactant will not result in adequate disruption of the supramolecular complexes which make up the membrane, thereby producing components larger than can be readily fractionated and causing the fractionation pattern to appear smeared. Addition of excess surfactant will disrupt intracomplex as well as intercomplex associations and release the chlorophylls and carotenoids from the binding protein because the pigments in the PPCs are non-covalently bound to the apoprotein. Thus it is necessary to determine empirically the amount of surfactant with which to treat the samples and to have present during fractionation. In fact, fractionation of PPCs by gel electrophoresis is sensitive to a large number of such factors which markedly affect the resulting fractionation pattern. The above statement is not meant to discourage researchers from attempting analysis of the photosynthetic PPCs, but rather to indicate that, although simple, the techniques are not trivial.

2.1 'Free' Pigment

Fractionation of PPCs by solubilisation and gel electrophoresis usually resolves multiple bands which are green and do not need to be treated further or stained for visualisation. Because the complexes are not fully denatured, the relationship between distance migrated and actual size is not necessarily predictable. However, it is usually safe to presume that the faster migrating complexes are smaller than those migrating more slowly. In almost all electrophoretic fractionation systems for the PPCs, the most rapidly migrating green band is not a PPC, but rather is generally termed 'free pigment'. Free pigment is probably complexed with surfactant in the form of mixed micelles and is assumed to represent chlorophyll and carotenoid that was formerly associated with protein, but which was released when the PPC was denatured during the solubilisation or fractionation. The laboratories developing electrophoretic fractionation procedures for PPCs generally have one of two different philosophies for what constitutes an improved procedure: an increase in the number of complexes resolved or a decrease in the amount of 'free pigment' generated. My own bias is for as little free pigment as possible and that will be reflected in the choice of fractionation systems to be presented in this chapter.

2.2 Oligomers

When a number of pigment-protein complexes is fractionated by gel electrophoresis, each may not be a distinct and unique complex. It appears that a number of the complexes studied may be oligomeric forms of smaller PPCs. However, it is not always clear whether such larger complexes are hetero-oligomers or homo-oligomers, that is oligomers made up of identical or different subunits (2,7). Examples of this are the light-harvesting LHCP1, LHCP2 and LHCP3 complexes and AB-1, AB-2 and AB-3 complexes fractionated by the Anderson system and the Markwell and Reinman (MARS) system, respectively. When attempts are made to quantitate the amount of each complex, this oligomeric nature of the various complexes can cause some complications. Seemingly minor differences in the solubilisation procedure can cause a change in the relative proportions of two or three oligomerically related PPCs, or even cause one or two not to be resolved. This possibility must be kept in mind when attempts are made to analyse the PPC composition of the photosynthetic membrane from fractionation patterns.

2.3 **Spectral Artefacts**

Absorption spectra are the most commonly determined characteristics of fractionated PPCs (Chapter 4). While these spectra can be useful, it must be realised that their relationship to the *in vivo* absorbance of these complexes is unknown. The chlorophyll chromophore is very sensitive to its molecular environment. Changes in the polarity of the milieu immediately surrounding the chlorophyll molecule will be reflected in changes in the wavelength of light absorbed, particularly in the red region (650 – 700 nm) of the spectrum, and changes in the extinction coefficient for absorption (8). Even sub-solubilising concentrations of surfactants cause marked effects on the absorption spectrum of the thylakoid membrane (9), and the concentrations used in solubilisation prior to fractionation of PPCs may disrupt both intercomplex and intracomplex interactions of the pigment molecules and proteins. Thus the absorption spectra of isolated PPCs cannot be used to correlate specific PPCs with any given chlorophyll spectral form in the intact thylakoid membrane. Absorption spectra can, however, provide information as to the presence and ratios of chlorophylls *a* and *b* as well as providing a qualitative estimate of the amount of phaeophytin produced during the fractionation.

3. APPLICATIONS OF THE TECHNIQUE

3.1 **Qualitative Determination of PPC Composition**

One of the more useful and easy to interpret applications of PPC fractionation is qualitatively to determine which PPCs are present or absent from a given plant. This has been used to examine the composition of the photosynthetic apparatus in mutants either totally lacking (10 – 13) or partially reduced (11,14) in chlorophyll *b*, a non-essential pigment which functions in the photosynthetic antenna. The mutants are lacking or have reduced amounts of specific PPCs which contain both chlorophylls *a* and *b*. Such data have established that chlorophyll *a/b*-protein complexes are the major light-harvesting components in the photosynthetic membranes of normal plants. Other mutants that have been examined using this technique include ones deficient in the electron transport activities associated with photosystem I (11) or photosystem II (15). These mutants were lacking specific pigment protein complexes which contained only chlorophyll *a*. Such data helped corroborate the suspicion that these PPCs represented the core of their respective photosystems.

Development of the photosynthetic apparatus is a second area in which qualitative studies on the presence or absence of the various PPCs have been employed. This has included the greening of either etiolated tissues (16) or of plants grown under an intermittent light regime (17,18). The former lack the photosynthetic PPCs and their appearance as a function of time in the light can be followed. The latter initially contain some of the chlorophyll *a*-containing PPCs associated with either photosystem I or II activity and the appearance of antenna components can be followed as greening occurs.

3.2 **Quantitative Determination of PPC Composition**

Perhaps the most desired analysis of PPCs is to determine quantitatively the proportion of pigment in each of the fractionated PPCs and thereby have some understanding of the proportion of each of these components within the photosynthetic apparatus. Quantitative analyses of PPC composition have been used to compare the efficacy of different fractionation systems, to examine the photosynthetic apparatus within the various domains of the thylakoid membrane, to observe changes in PPC abundance during development of the thylakoid membrane and to observe the effect of environment and environmental stresses on the photosynthetic apparatus. While such studies can be carried out and provide useful and needed information (16,18–20), several potential problems must be considered which could affect interpretation of the results. These problems include the method of quantitation, the amount of free chlorophyll, the presence of oligomers and the stoichiometry of surfactant to membrane in the solubilisation process.

The method used to quantitate the amount of pigment in the various fractionated PPCs will have a direct bearing on the significance of the resulting data. The PPCs can contain both chlorophyll *a* and *b*, and the two pigments have different absorption spectra and different extinction coefficients. Furthermore there are a number of spectral forms of pigments within any given PPC, causing a broadening of the composite absorption in the red region of the visible spectrum – the most useful region for these studies due to the lack of interfering absorption by other thylakoid components. The usual method of PPC quantitation is to record the absorbance of the gel as it is moved through the light beam of a spectrophotometer. This can be done at a wavelength that is absorbed almost exclusively by chlorophyll *a* (e.g. 675 nm) and will give an estimate of the proportion of that pigment in each of the fractionated PPCs. The problem with this method is that it ignores the presence of chlorophyll *b*, a significant component in some of the PPCs. To correct for this failing, some laboratories scan the gel containing the fractionated PPCs twice, first at a wavelength appropriate for chlorophyll *a* (e.g. 675 nm) and the second time at a wavelength which is absorbed predominantly by chlorophyll *b* (e.g. 650 nm). The relative absorbances at each wavelength are averaged and used to quantitate the proportion of total chlorophyll in each PPC. Unfortunately, chlorophyll *a* has a greater extinction coefficient than chlorophyll *b*, and this method tends to overestimate the total amount of pigment in the chlorophyll *b*-containing PPCs. A third method which should circumvent the deficiencies of the first two is to excise pieces of gel containing each of the PPCs and extract the pigments into an appropriate solvent for spectrophotometric analysis (Chapter 4). Problems in obtaining enough of all the PPCs to produce sufficient absorbance for accurate analysis, as well as problems in complete extraction and recovery which adversely affect precision, make this method more time consuming and no better than the previous two.

4. DIFFERENT TYPES OF FRACTIONATION SYSTEMS

Unfortunately, no one electrophoretic fractionation system serves all purposes. Consequently, there is a number of such systems (and modifications) currently being used in studies on photosynthetic PPCs. Several different fractionation systems will be presented in sufficient detail to allow most researchers with at least some biochemistry experience to obtain results after going through the procedure only a few times. The primary difference between electrophoretic fractionation methods is whether the buffer system in each is continuous or discontinuous. Fractionation systems employing continuous buffer systems can theoretically subject samples to less harsh treatment during the electrophoresis than discontinuous systems, but have the disadvantage that samples do not focus during fractionation and thus smaller sample volumes must be loaded onto the gels. Consideration of denaturation or alteration of PPCs during electrophoretic fractionation is not a trivial matter. The same preparation of thylakoid membranes solubilised identically with sodium dodecylsulphate (SDS) can yield an average of approximately 2, 10 or 50% free pigment, depending on the fractionation system used (5,20).

4.1 Common Procedures

4.1.1 *Thylakoid Membrane Isolation*

The procedures used in the isolation of thylakoid membranes for studies on PPC fractionation vary somewhat between laboratories, but are relatively similar. The following is a procedure that can be generally adapted and used with most fractionation systems.

(i) Harvest leaves and remove any large mid-ribs.

(ii) Coarsely cut or chop the leaves and place in ice-cold buffer consisting of 50 mM sodium Tricine (pH 8.0), 0.4 M sucrose, 10 mM NaCl and 5 mM $MgSO_4$. Grind the leaves in a chilled mortar with a pestle or mechanically disrupt them in a device such as a Polytron homogeniser or Waring blender. Two 5-sec periods of mechanical grinding are usually sufficient.

(iii) Filter the resulting homogenate through some medium such as muslin, cheese cloth, controlled pore nylon mesh or Miracloth (a siliconised paper) to remove large debris.

(iv) Centrifuge the filtrate at 3000 *g* for 1 min to pellet a fraction enriched in intact chloroplasts and photosynthetic membranes.

(v) Discard the supernatant fraction and resuspend the pellet in cold wash buffer (25 mM Na Tricine, 1 mM EDTA, pH 8.0) with the aid of a vortex mixer. Disrupt thoroughly with approximately five cycles in a pre-chilled glass tissue grinder (e.g. Ten Broek or Potter Elvehjem type).

(vi) Centrifuge the washed membranes at 30 000 *g* for 10 min; wash and centrifuge the resulting pellet once more.

The last pellet can be used immediately for surfactant extraction and PPC fractionation by the Thornber, MARS or Deriphat 160 systems (see below) or it can be washed in 50 mM Na Tricine (pH 8.0) and used immediately or frozen in liquid nitrogen for later fractionation by the Anderson system. Throughout the above isolation, it is important to keep the solutions and membranes cold at all times, to pre-chill the glass homogenisers and to work in a rapid manner.

For most samples, the method of Arnon (21) using 80% acetone to extract pigment works well for determining chlorophyll content (Chapter 4). An aliquot of the membranes, usually between 10 and 100 μl, is placed into a 12 or 15 ml glass conical centrifuge tube. Sufficient water is added to bring the total volume to 1.0 ml. Then 4.0 ml of acetone is added, the tube mixed and spun at approximately 3000 *g* in a clinical centrifuge. The absorbance of the solution at 663 and 645 nm is determined relative to a control tube containing 1 ml of water and 4 ml of acetone. Personal experience indicates that the calculation of chlorophyll concentrations is often confusing at first for new researchers and is a frequent cause of bad fractionation patterns. I have found it desirable to equip my laboratory with a BASIC microcomputer program (*Table 1*)

Table 1. BASIC Microcomputer Program for Chlorophyll Concentration.

```
10 CLS
20 PRINT "*****************************************************************"
30 PRINT "     PROGRAM FOR DETERMINATION OF CHLOROPHYLL CONCENTRATION       "
40 PRINT "*****************************************************************"
50 PRINT
60 INPUT "ABSORBANCE AT 663 nm = ? ", A663: PRINT
70 INPUT "ABSORBANCE AT 645 nm = ? ", A645: PRINT
80 INPUT "HOW MANY MILLILITERS OF 80% ACETONE? ", VOLUME: PRINT
90 INPUT "HOW MANY MICROLITERS OF SAMPLE? ", ALIQUOT: PRINT
100 INPUT "ALIQUOT WAS FROM HOW MANY MILLILITERS OF MEMBRANES? ", MEMBRANES
110 REM ***** CALCULATION BASED ON MACKINNEY-ARNON EQUATIONS *****
120 CHLA = ((12.7 * A663)-(2.6 * A645)) * VOLUME * MEMBRANES / ALIQUOT
130 CHLB = ((22.9 * A645)-(4.68 * A663)) * VOLUME * MEMBRANES / ALIQUOT
140 TOTALCHL = CHLA + CHLB
150 RATIO = CHLA/CHLB: PRINT
160 PRINT "MILLIGRAMS CHLOROPHYLL a + b IN SAMPLE = "; TOTALCHL: PRINT
170 PRINT "RATIO CHLOROPHYLL a/b = "; RATIO: PRINT
180 IF RATIO <2.5 THEN PRINT "CHL a/b LESS THAN NORMAL, POSSIBLE ERROR!!"
190 IF RATIO >3.5 THEN PRINT "CHL a/b GREATER THAN NORMAL, POSSIBLE ERROR!!"
200 PRINT: INPUT "ANOTHER SET OF CALCULATIONS? (Y/N) ", Z$
210 IF Z$ = "Y" OR Z$ = "y" THEN GOTO 50
220 IF Z$ = "N" OR Z$ = "n" THEN END
230 PRINT "PLEASE ANSWER 'Y' OR 'N' ": GOTO 200
```

which aids in the calculation and warns if the absorbance values seem inappropriate for most normal plants.

4.1.2 *Surfactant Treatment of Thylakoid Membranes*

Aside from accurate measurements of chlorophyll, the most important aspect of the solubilisation process is that it be carried out in the cold. Solutions of SDS to be used for solubilisation of PPCs should be warmed to approximately 30°C and a small volume put on ice approximately 5 min prior to the extraction. Following addition of the surfactant, the thylakoid-containing solution should be mixed thoroughly. For sample volumes large enough (>500 μl) it is recommended that the sample be further mixed with an appropriately sized tissue grinder. For the Anderson fractionation system, the sample is loaded immediately onto the gel; for the other systems the sample is first centrifuged to remove any debris or starch (which may clog the pores in the gel and cause the sample to smear) and then immediately loaded onto the gel. A microsyringe (25 or 50 μl) is generally the best device for loading samples.

4.1.3 *Casting Polyacrylamide Gels*

For the theory behind acrylamide gels and their polymerisation the reader should consult the first volume in this series (6). The major problems encountered in casting gels are proper sealing of the plates for slab gels and consistent polymerisation. The author has found the use of three spacers and molten agarose (6) to provide the most satisfactory method for sealing slab gel plates. With regard to consistent polymerisation, mild warming has been found to work well. Solutions containing all the gel components except the ammonium persulphate are warmed in a water bath to approximately 30°C. Ammonium persulphate can then be added, the solution mixed by gentle swirling and immediately poured into tubes or slab gel plates. It is not recommended that organic solvent such as isopropanol be used to layer onto the gels during the polymerisation to aid in forming a flat upper gel surface. When this was done with some of the fractionation systems mentioned in this chapter, the amount of free chlorophyll increased and the number of apparent PPCs increased dramatically. It is strongly suspected that these results were artefacts caused by residual organic solvent left at the top of the gel.

4.2 Common Reagents and Solutions

4.2.1 *Acrylamide*

Acrylamide and N,N′-methylenebisacrylamide should be of high purity for PPC fractionation. Suppliers whose acrylamide has proved satisfactory include Bio-Rad Laboratories, Bethesda Research Laboratories, Eastman Organic Chemicals and Miles Laboratories. Because both acrylamide and N,N′-methylenebisacrylamide are neurotoxins, purchase of high quality reagents also obviates the somewhat hazardous task of recrystallisation required with less pure grades. Safety precautions should be strictly adhered to when working with acrylamide, either in solid form or in solutions. Acrylamide solutions are generally prepared as 20 or 30% solutions with the appropriate amount of N,N′-methylenebisacrylamide added as cross-linking component. For a 20 or 30% solution add 40 or 60 g, respectively, of acrylamide with the appropriate amount of N,N′-methylenebisacrylamide to 140 ml of water and stir until dissolved. This is an endothermic process and may take 30 min. Once dissolved, add water to a final volume of 200 ml. Filter through paper (Whatman No. 1) and store in a dark bottle in the cold. Solutions can be stored in the cold for as long as 3 months without causing noticeable loss of resolution.

4.2.2 *N,N,N′,N′-Tetramethylethylenediamine (TEMED)*

This is used as supplied. It should be a colourless liquid. TEMED is a carcinogen and should be used in a fume hood with appropriate precautions. When your gels do not polymerise it is most likely that the TEMED has become contaminated. It is generally good practice to purchase this reagent in small amounts, label with the date opened, and replace with fresh reagent each year.

4.2.3 *Ammonium Persulphate*

Prepared as a 10% solution. Dissolve 0.2 g in 2 ml of water. This reagent should be prepared fresh daily.

4.2.4 *Surfactants*

Surfactants should be as pure as possible. Non-ionic surfactants are generally a mixture of molecules with a common hydrophobic moiety and different lengths of some hydrophilic polymer. While sold with a stated average length of hydrophilic polymer, the solution is not homogeneous and this will affect the molecular properties of the surfactant solution. These properties can include critical micelle concentration and cloud point. Thus different lots of the same surfactant, or two surfactants differing only in the statistical abundance of the various polymeric chain lengths which make up the same average length (e.g. Triton X-100 and Nonidet P-40), can have different effects on PPC fractionation patterns. The same is true for ionic surfactants such as SDS. The source of surfactant is one of the more likely areas of PPC fractionation that will cause bad or inconsistent results. It is generally advisable for the researcher just initiating work in this field to purchase small amounts of a surfactant such as SDS from several different suppliers. The best source of surfactant must be determined empirically. Once a good source is identified, it is wise to see if the supplier can sell a larger quantity from the same production lot.

4.3 **Continuous Fractionation Systems**

Two different electrophoretic fractionation systems using continuous buffer systems will be described. These systems resolve increasingly larger and more complex PPCs and the reason for the use of each will be different.

4.3.1 *Thornber System*

The system of Thornber (13,22) resolves two PPCs in addition to a zone of free chlorophyll (*Figure 1*). These PPCs are termed CP-I and CP-II. CP-I contains only chlorophyll *a* and is derived from photosystem I. CP-II contains chlorophylls *a* and *b* in ratios of approximately 1:1, represents approximately 45% of the chlorophyll in the sample, and is derived from LHC-2, the light-harvesting complex of photosystem II. An oligomeric form of CP-II, migrating between CP-I and CP-II, has also been reported (23,24). The free chlorophyll zone contains approximately 45% of the chloro-

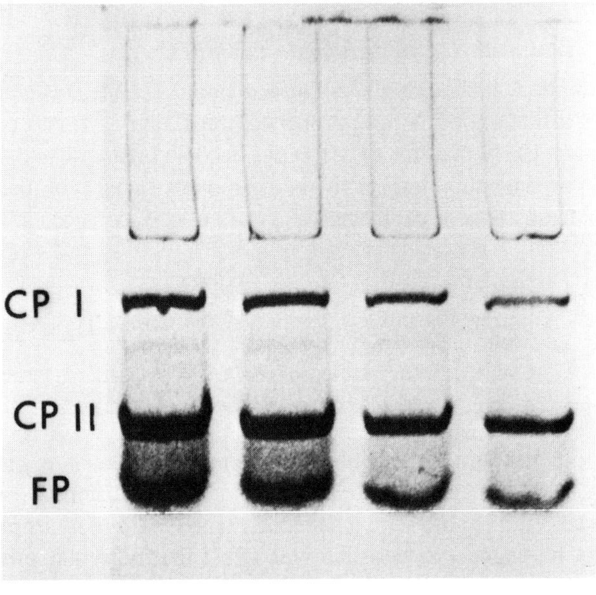

Fig. 1. Pigment-protein complexes fractionated on the Thornber electrophoretic system. Decreasing amounts of surfactant-treated thylakoid membranes (30, 20, 10 and 5 μl) were loaded from left to right onto the gel and fractionated as described in the text. This is the pattern resolved immediately after fractionation without staining or other visualisation techniques required. Two PPCs, CP-I and CP-II, and a zone of free pigment (FP) are visible. An oligomeric form of the CP-II complex is barely visible migrating between CP-I and CP-II. The wells formed into the gel for loading the samples are visible above the fractionated PPCs.

Table 2. Media and Gel Preparation for the Thornber System for Separating Pigment-protein Complexes.

Extraction Buffer (50 mM Tris-Cl,10% glycerol, 1% SDS, pH 8.0)
1. Dissolve 0.60 g of Tris base, 10 ml of glycerol and 1 g of SDS in water.
2. Adjust pH to 8.0 with HCl and add water to 100 ml.

20% Acrylamide
3. Dissolve 40 g of acrylamide and 1.2 g of N,N′-methylenebisacrylamide in sufficient water to make 200 ml of solution.

4-fold Concentrated Buffer (200 mM Tris-Cl, 0.4% SDS, pH 8.0)
4. Dissolve 24.2 g of Tris base and 4 g of SDS in water.
5. Adjust pH to 8.0 with HCl and dilute with water to 1 l.

Reservoir Buffer
6. Dilute one part of concentrated buffer in three parts water.

To prepare gel
7. Mix 14 ml of acrylamide solution, 10 ml of concentrated buffer solution, 20 μl of TEMED and 15.5 ml of water.
8. Add 0.5 ml of 10% ammonium persulphate, mix and pour. Gels are ready to use 1 h after casting.

phyll applied to the gel. The primary advantage of the Thornber fractionation is that it is very likely to work on the first attempt. The more often that PPCs are fractionated in a laboratory, the more likely that the fractionation patterns will be acceptable. This fractionation system usually produces the expected pattern and prevents frustration in initial experiments with PPCs.

(i) *Reagents and procedures for the Thornber system.* Recipes are shown in *Table 2*.

(ii) *Extraction procedure.* To pelleted thylakoid membranes add 1.0 ml of extraction buffer for each milligram of total chlorophyll. Resuspend the pellet and treat with a glass homogeniser or tissue grinder to ensure complete solubilisation. Centrifuge at 40 000 *g* for 5 min. Samples of 5 – 30 μl should be loaded onto the gels.

(iii) *Electrophoresis.* Run the fractionation at approximately 10 V per cm length of gel. Fractionation should be optimal in approximately 1 h.

4.3.2 *The Markwell and Reinman System (MARS)*

The MARS system (20) resolves four PPCs and the amount of free chlorophyll is reduced to approximately 15% or less. The major difference from the Thornber system is the use of a Tris-glycine buffer system, rather then Tris-Cl. The four PPCs (*Figure 2*) are termed A-1, AB-1, AB-2 and AB-3. A-1 is derived from the core of photosystem I and contains almost no chlorophyll *b*. AB-1, AB-2 and AB-3 contain chlorophyll *b* in addition to chlorophyll *a* and are thought to be derived from LHC-2. AB-3 also appears to co-migrate with and obscure a PPC termed A-2 (11,12) which is derived from photosystem II and is only observed with some difficulty using the MARS system (11). The A-2 complex can be seen clearly in mutants lacking chlorophyll *b* (*Figure 3*), also demonstrating the use of electrophoretic fractionation to analyse mutations affecting photosynthetic pigment composition.

The major advantage of this system is that it allows recovery of a large proportion of the chlorophyll with the PPCs and has a high probability of providing consistent

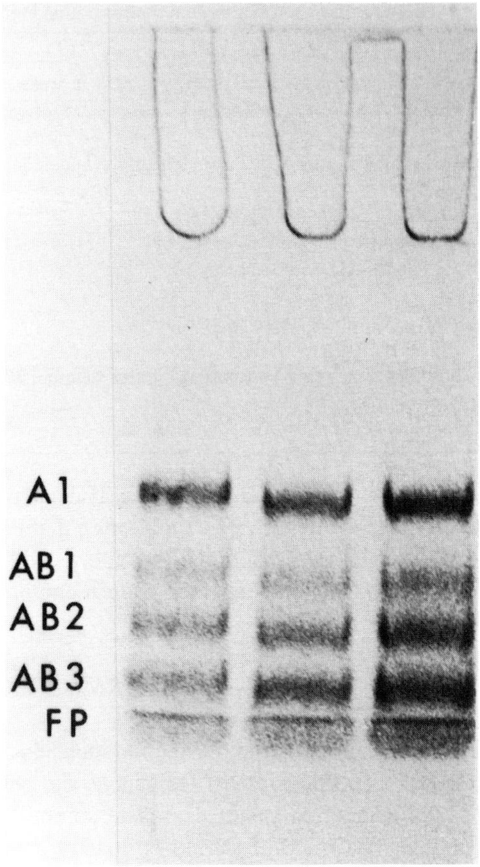

Figure 2. Pigment-protein complexes fractionated from surfactant-solubilised thylakoid membranes by the MARS system. Increasing amounts of sample (5, 10 and 20 μl) were applied from left to right. One chlorophyll *a* (A-1) and three chlorophyll *a*+*b* (AB-1, AB-2 and AB-3) PPCs and a zone of free pigment are resolved. The sample wells are visible above the fractionated PPCs.

PPC fractionation within the first few attempts. We use the MARS system as an undergraduate biochemistry laboratory exercise at the University of Nebraska and rarely have an unsuccessful fractionation.

(i) *Reagents for MARS fractionation system.* Recipes are shown in *Table 3*.

(ii) *Extraction procedure.* To pelleted thylakoid membranes add 1.0 ml of chilled extraction buffer per milligram of chlorophyll in the sample. Keeping the solution cold, resuspend the pellet and treat with a glass homogeniser. Centrifuge at 40 000 *g* for 5 min. Load 5 – 25 μl of the supernatant fraction onto the gel.

(iii) *Electrophoresis.* Run the fractionation at room temperature with approximately 10 V per cm of gel. Fractionation should be optimal in 45 – 60 min.

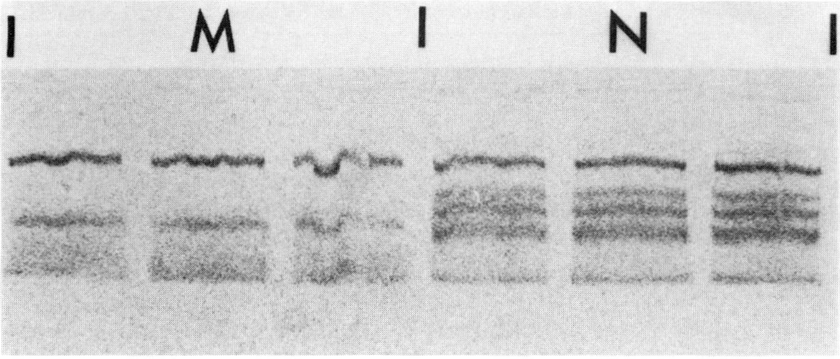

Figure 3. Fractionation of surfactant-solubilised thylakoid membranes isolated from normal (N) and chlorophyll *b*-lacking mutants (M) of sweetclover (*Melilotus alba*) using the MARS system. The normal membranes give rise to the expected fractionation pattern (cf. *Figure 2*) whereas the mutant membranes show an altered pattern. The band labelled A-2 is thought to be present in the normal thylakoids, but is usually obscured by co-migration with AB-3. The A-2 PPC, which is probably similar to the CPa complex fractionated with the Anderson system (see *Figure 4*), contains only chlorophyll *a* and is derived from photosystem II. The uneven migration of the bands in the third column from the left is due to an air bubble trapped under the sample well-forming comb during polymerisation of the acrylamide gel.

Table 3. Media and Gel Preparation for the Markwell and Reinman System (MARS) for Separating Pigment-protein Complexes.

Extraction Buffer (6.2 mM Tris, 48 mM Glycine, 10% Glycerol, 1% SDS)

1. Dissolve 0.19 g of Tris, 0.9 g of glycine, 25 ml of glycerol and 2.5 g of SDS in water and dilute to 250 ml total volume.

20% Acrylamide

2. Dissolve 40 g of acrylamide and 1.06 g of N,N′-methylenebisacrylamide in water and dilute to 200 ml.

10-fold Concentrated MARS buffer (62 mM Tris, 480 mM Glycine, 1% SDS, pH ~8.3)

3. Dissolve 7.5 g of Tris base, 36 g of glycine and 10 g of SDS in water, dilute to a final volume of 1 l. The pH of the solution should be approximately 8.3 without adjustment.

Reservoir buffer

4. Dilute 1 part MARS buffer with 9 parts water.

To prepare gel

5. Mix 10 ml of acrylamide, 4 ml of concentrated buffer solution, 25.5 ml of water and 20 μl of TEMED.
6. Add 0.5 ml of 10% ammonium persulphate, mix and pour. Gels are ready to use 1 h after casting.

4.4 **Discontinuous Fractionation Systems**

Two types of discontinuous fractionation systems will be described. The first, the Deriphat 160 fractionation system, is discontinuous only in that the buffer concentrations in the gel and reservoirs are 2-fold different. This system yields less free chlorophyll than the other systems, but larger and less simple PPCs. The second system, the Anderson system, has the potential to fractionate a larger number of PPCs than the other systems, but may be more difficult to get to run correctly on the initial attempts. This system is based on the discontinuous buffer system of Neville (25) and uses a stacking gel which focusses the sample during the initial phase of fractionation and permits utilisation of larger sample volumes.

4.4.1 *The Deriphat 160 Fractionation System*

The Deriphat 160 system (5) usually resolves four PPCs, termed ‘N’, ‘O’, ‘P’ and

Table 4. Media and Gel Preparation for the Deriphat 160 System for Separating Pigment-protein Complexes.

Extraction Buffer (6.2 mM Tris, 48 mM Glycine, 10% Glycerol, 1% SDS)
1. Dissolve 0.19 g of Tris, 0.9 g of glycine, 25 ml of glycerol and 2.5 g of SDS in water and dilute to 250 ml total volume.

20% Acrylamide
2. Dissolve 40 g of acrylamide and 2 g of N,N′-methylenebisacrylamide in water and dilute to 200 ml.

4-fold Concentrated Gel Buffer (24.8 mM Tris, 192 mM Glycine, 0.4% Deriphat 160, pH ~8.3)
3. Dissolve 0.75 g of Tris base, 3.6 g of glycine and 1 g of Deriphat 160 in water and dilute to a final volume of 250 ml.

10-fold Concentrated Reservoir Buffer (124 mM Tris, 960 mM Glycine, 2% Deriphat 160, pH ~8.3)
4. Dissolve 15 g of Tris base, 72 g of glycine and 20 g of Deriphat 160 in water and dilute to a final volume of 1 l. Dilute one part buffer with 9 parts water before use.

To prepare gel
5. Mix 10 ml of 20% acrylamide, 10 ml of gel buffer, 19.5 ml of water and 20 μl of TEMED.
6. Add 0.5 ml of 10% ammonium persulphate, mix and pour. It is normal for the solution to become somewhat cloudy upon addition of ammonium persulphate. Gels are ready 1 h after casting.

'Q'. N and O both contain P700 and are derived from photosystem I. Studies with tobacco suggest they differ in their composition of the two chlorophyll *b*-containing antenna complexes, LHC-1 being more prevalent in the former whereas AB-0 is more abundant in the latter. The P complex has a chlorophyll *a*/*b* ratio of approximately 1.3 and represents an oligomeric form of the light-harvesting complex. The P complex will occasionally split during fractionation into two or three smaller PPCs containing both chlorophylls *a* and *b*. The Q complex appears to be derived from photosystem II. The amount of free chlorophyll is usually less than 5% of that applied to the gel, and is frequently undetectable. Since this system uses the same solubilisation buffer as the MARS system, the decrease in free chlorophyll implies that most free chlorophyll observed in that system is probably generated by the electrophoretic fractionation rather than by the solubilisation.

Due to the supramolecular nature of the resolved PPCs, it is not recommended that this fractionation system be used to examine the PPC composition of the photosynthetic apparatus. Rather, it appears to offer some advantages over the other systems in studies on the molecular architecture and organisation of the PPCs.

(i) *Reagents and procedures for Deriphat 160 fractionation system.* Recipes are shown in *Table 4*. Note that Deriphat 160 (N-lauryl-β-iminodiproprionate) normally contains 10–15% of the monocarboxyl form of the surfactant. For consistent results, it may be advisable to purchase purified Deriphat 160 which is predominantly dicarboxyl form. The purified surfactant is usually somewhat basic and the above buffers may need to be adjusted so that the final pH is approximately 8.3; a cation exchange resin (H^+-form) is used for this purpose. Deriphat 160, purified Deriphat 160 and Deriphat 160C (a 30% solution of Deriphat 160) may be purchased from McKerson Chemical Distribution Co., Minneapolis, MN 55434, USA.

(ii) *Extraction procedure.* This is identical to the extraction procedure used for the MARS fractionation system. Between 5 and 20 μl of the supernatant fraction should be loaded on the gels. Greater volumes appear to increase the amount of free chlorophyll. As

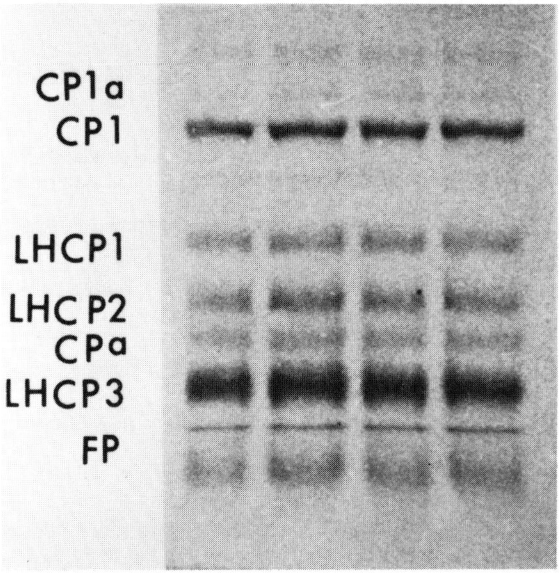

Figure 4. Fractionation of surfactant-solubilised thylakoid membranes by the Anderson system. Identical amounts of sample were loaded onto the gel and fractionated as described in the text. Six PPCs and two zones of free pigment (one sharp and the other diffuse) are resolved. The complexes CP1 and CP1a (difficult to see) are derived from photosystem I whereas CPa is derived from photosystem II; these bands contain chlorophyll *a*. The bands designated LHCP contain both chlorophylls *a* and *b*.

this fractionation will often result in no apparent free pigment, it is sometimes difficult to discern whether the fastest migrating green zone during fractionation is a PPC or free chlorophyll. Therefore, it is advisable to heat-treat a small portion of the extracted supernatant (60 – 80°C for a few minutes) to generate a large amount of free chlorophyll to use as reference.

(iii) *Electrophoresis.* The gel should be run at approximately 7.5 V per cm of length. Fractionation should be optimal in approximately 45 min.

4.4.2 *The Anderson Fractionation System*

The Anderson electrophoretic system has evolved somewhat since it was first introduced (10). The recipe described here was provided by Saeid Nourizadeh and Philip Thornber at the University of California-Los Angeles and is based on a modification (26) of the original. This fractionation system will generally resolve six PPCs and a zone of free pigment comprising 10 – 15% of the applied chlorophyll (*Figure 4*). Two of the PPCs, CP1 and CP1a are derived from photosystem I. CP1a appears to migrate more slowly than CP1 because it additionally contains LHC-1, a chlorophyll *b*-containing antenna complex (27). The PPC CPa is derived from photosystem II and contains little, if any, chlorophyll *b*. The three complexes which contain the majority of the chlorophyll *b*, LHCP-1, LHCP-2 and LHCP-3, are derived from the LHC-2. It is also possible to resolve additional PPCs which migrate more slowly than CP1a and are also

Table 5. Media and Gel Preparation for the Anderson System for Separating Pigment-protein Complexes.

Solubilisation Buffer [300 mM Tris-Cl, 13% (v/v) Glycerol, 0.375% (w/v) SDS, pH 8.8]
1. To 75 ml of water add 13 ml of glycerol, 3.63 g of Tris base and 0.375 g of SDS.
2. Adjust pH to 8.8 with HCl and dilute to a final volume of 100 ml.

30% Acrylamide
3. Dissolve 58.4 g of acrylamide and 1.6 g of N,N′-methylenebisacrylamide in water and dilute to 200 ml.

Lower Reservoir Buffer and 2-fold Concentrated Lower Gel Buffer (430 mM Tris-Cl, pH 9.35)
4. Dissolve 104 g of Tris base in 1800 ml of water.
5. Adjust pH to 9.35 with HCl and dilute to a final volume of 2 l.

2-fold Concentrated Upper Gel Buffer (112 mM Tris-SO_4, pH 6.14)
6. Dissolve 2.72 g of Tris base in water, adjust to pH 6.14 with H_2SO_4 and dilute to a final volume of 200 ml.

Upper Reservoir Buffer (41 mM Tris-borate, pH 8.64)
7. Dissolve 9.93 g of Tris base in 1.8 l of water.
8. Adjust pH to 8.64 with a saturated solution of boric acid and dilute to a final volume of 2 l.

To prepare gel
9. Mix 9.3 ml of acrylamide, 7.9 ml of water, 17.5 ml of lower gel buffer and 17.5 μl of TEMED.
10. Add 0.35 ml of 10% ammonium persulphate and mix.
11. Pour solution between glass plates to a sufficient depth so that 1 cm of space will remain between the upper surface of the gel and the bottom of the well-forming comb.
12. Gently layer water on top of the acrylamide solution and let polymerise for 30 – 60 min.
13. When the lower gel has polymerised, pour off the water and prepare the upper gel solution: 0.96 ml of acrylamide, 3.6 ml of upper gel buffer, 2.64 ml of water, 72 μl of 10% SDS, 3.6 μl of TEMED and 72 μl of 10% ammonium persulphate.
14. Pour this solution on top of the lower gel, insert the sample well-forming comb and let polymerise.
15. After 30 min, the comb can be removed and the upper reservoir buffer pipetted into the sample wells.
16. Allow the gel to continue polymerisation for an additional hour and then place into a cold room or refrigerator and chill to approximately 4°C. Reservoir buffers should also be chilled, but if allowed to remain cold for too long a period the SDS will become insoluble.

thought to be derived from photosystem I (19).

It has been the author's experience that this fractionation system is more sensitive than those described above to suboptimal conditions or reagents causing poor resolution of the constituent PPCs in a sample. Careful attention must be paid to the sources of all the reagents, particularly the SDS, as well as to the conditions and temperatures used in the solubilisation and electrophoresis. This is not meant to discourage use of this system. The Anderson system has been the most widely used in quantitative studies on the composition of the photosynthetic apparatus and, when properly done, it results in excellent fractionation of the core components of photosystem I and II. In this regard it may be superior to the other systems presented in this report. The loss of resolution caused by different sources of reagents and lack of careful temperature control can be overcome by patience and an empirical approach.

(i) *Reagents and procedures for Anderson fractionation system.* Recipes are shown in *Table 5.*

(ii) *Extraction procedure.* Membranes can be used either freshly prepared or following storage in liquid nitrogen. For each milligram of chlorophyll in the membrane pellet, 2 ml of solubilisation buffer are added. The tube is vigorously mixed and samples of 40 – 60 μl should be loaded directly into the sample wells.

(iii) *Electrophoresis.* Fractionation in the cold is carried out initially with sufficient voltage to produce approximately 1 mA per cm of gel length. When the sample has migrated into the lower gel, the voltage is increased to produce approximately 2 mA per cm of gel length. Fractionation should be complete in approximately 3–4 h. It is important that the gel and reservoir buffers be completely chilled prior to initiating fractionation as warming the system by even 5°C can dramatically reduce resolution of the larger and oligomeric PPCs. The original description of this procedure recommends a pre-electrophoresis of the gel prior to loading the sample; we have not found this step to be necessary or critical for good resolution.

4.5 Other Fractionation Systems

A variety of other electrophoretic systems has been employed in efforts to improve the resolution of fractionated PPCs. Of these, two will be mentioned. The first is the system of Delepelaire and Chua (28) using the surfactant lithium dodecylsulphate, which is more soluble in the cold than the corresponding sodium salt. This system is able to resolve two PPCs from photosystem II with different apoproteins and pigment contents. A recent modification of this system (15) using a different solubilisation process shows that the number of PPCs resolved can be increased and demonstrates how this system can be used in conjunction with plant mutants to identify PPCs having specific functions. The major negative aspect of this fractionation system is that the amount of free chlorophyll generated is significantly larger than with the MARS or Anderson systems.

The second fractionation system was developed by Siefermann-Harms and Ninnemann (29) and involves fractionation in polyacrylamide gels by isoelectric focussing. A number of PPCs was resolved and the chlorophyll *b*-containing complexes were well resolved from complexes derived from photosystems I and II. The limitation to resolution would appear to be the relatively similar pI values (between pI 4 and 5) of all the PPCs. This system would appear to be potentially quite useful but has been little used.

5. EQUIPMENT CONSIDERATIONS

5.1 Tube Versus Slab Gels

There is no one proper type of electrophoretic apparatus for PPC fractionation studies. These studies were all initially carried out with tube gel systems, and then in about 1977 slab gels became more common. The general advantages and disadvantages of each have been previously discussed (6). The major advantage of tube gels with respect to PPC fractionation is that electrophoresis of the same sample can be halted at various times to quantitate the proportion of total pigment in each PPC. This proportion may change as the fractionation proceeds, with pigment progressively lost from some of the complexes; not all of the PPCs are equally stable. The fact that most devices for adapting spectrophotometers for scanning gels are designed for use with tube gels may also be a consideration in selecting the type of apparatus to use. The major advantage of slab gels is that fractionation patterns resulting from different samples or treatments of the same sample can be directly compared. This can be a great advantage when attempting empirically to improve the solubilisation conditions for a fractionation system.

Tubes and slab gels should generally be small enough so that the total length of polyacrylamide is 10 cm or less. Longer gels do not appear greatly to facilitate maintaining the pigment with the PPCs and could be considered detrimental. The reason for this is probably that the voltages needed to obtain equivalent currents become excessive. Very small slab vertical gel electrophoresis units can be purchased from companies such as Hoefer Scientific Instruments. It may be desirable, however, to have a single electrophoresis apparatus that can serve multiple purposes. The author has found several companies (American BioNuclear, CBS Scientific and FMC Corporation) who sell an adjustable apparatus that can run the smaller gels for PPC fractionation and can have their upper reservoir chambers raised to run gels 25 cm or longer for polypeptide analyses. These units are available with Teflon combs, to facilitate formation of smooth sample wells, and a variety of spacer thicknesses. Spacers 1.5 or 2 mm thick usually provide the best results, but 3 and 5 mm spacers are useful for preparative work.

5.2 **Horizontal Gels**

For laboratories which routinely work with DNA electrophoresis and may not be equipped for protein fractionation on vertical slab gels, it is also possible to use horizontal electrophoresis units for this purpose. At least two companies (BioRad Laboratories and Bethesda Research Laboratories) can provide a plate which will fit on top of the gel casting deck and form a suitable cavity for formation of a horizontal slab using their smallest sized apparatus. Acrylamide will not gel unless the upper plate is present because oxygen acts as a chain terminator for the free radical-catalysed polymerisation. The author's laboratory has successfully used these horizontal minigels with the Thornber, MARS and Deriphat 160 fractionation systems. This type of apparatus cannot be used with the Anderson system since it requires the presence of an upper stacking gel.

5.3 **Power Supplies**

For the fractionation systems described in this report, an inexpensive power supply that provides constant voltage up to approximately 250 V would be sufficient. Since the Anderson fractionation system is run at constant current and takes several hours to separate the PPCs, it may be desirable to have a power supply that provides constant current by automatically increasing the voltage.

6. FURTHER ANALYSIS

It is often desirable to fractionate further or analyse PPCs once they are fractionated. Examples of such studies would be to fractionate supramolecular PPCs into their constituent monomeric units, to determine the absorption spectrum of a fractionated PPC, to measure the pigment composition of a PPC or to determine which polypeptides are present in any given PPC.

6.1 **PPC Excision**

The easiest form in which to use fractionated PPCs is to excise the portion of the gel containing the material of interest. Polyacrylamide gels generally do not cut or slice well, but chopping with a straight-edged implement is effective. This can be a razor blade, a long-bladed spatula, or even a metal or plastic ruler. Position the implement onto the surface of the gel such that it aligns with the desired position for the slice and press down firmly. This is most easily done on top of a light box so that the PPCs are more visible. Since there are a large number of colourless polypeptides which have migrated in the gel, try to take as little excess gel as possible. When excising two adjacent PPCs it is preferable also to cut a section of gel between the two complexes and discard it.

6.2 **PPC Elution From Gel Slices**

Electroelution of PPCs from excised gel slices is rapid and efficient, but an apparatus for performing this technique is not always available. The following procedure (23) requires no special equipment.

(i) Place the excised gel slices inside the barrel of a small plastic syringe.
(ii) Insert the plunger and extrude the gel slices out of the syringe barrel and through a piece of nylon fabric with a small ($\sim 100\ \mu m$) mesh size.
(iii) The extruded gel will form a sticky mass. Pick this up with a spatula, place again in the syringe and re-extrude.
(iv) Place the extruded gel material in a small beaker or flask and add appropriate buffer solution.
(v) Place the container in a cold room and either stir or shake it for several hours to facilitate diffusion of the complexes from the polyacrylamide. Because the rate of diffusion from the gel will be inversely related to the size of the PPC, very large complexes such as those originating from photosystem I will take longer to enter the buffer.
(vi) Centrifuge the slurry to remove the polyacrylamide and remove the supernatant containing the eluted PPCs for further analysis.

6.3 **Re-electrophoresis**

Excised gel slices containing PPCs can be placed in the sample wells of a second gel and re-electrophoresed on the same or a different fractionation system. This technique is useful in trying to establish whether two PPCs with similar properties are actually identical (30) or to disrupt larger PPCs further into their constituent monomeric complexes. This is demonstrated in *Figure 5* which shows re-electrophoresis of the chlorophyll *b*-containing PPCs from the MARS and Thornber systems. This type of analysis is used to establish that these complexes with similar properties are indeed different from one another. If the second electrophoresis is for the analysis of constituent polypeptides, the gel slice should be incubated in a denaturing solution prior to re-electrophoresis. The actual placing of the gel slices in the sample wells is best done with a small spatula and will take some practice. It is often convenient to make the second gel slightly thicker than the first (i.e. a 1.5 mm first and 2.0 mm second gel).

6.4 **Spectral Analysis**

Analysis of PPC absorption spectra can be done with gel slices. A semi-micro cuvette (1 cm path length accommodating $0.5-1$ ml volume) is filled two-thirds full with gel buffer that has been diluted to the same concentration as that present in the gel. Two

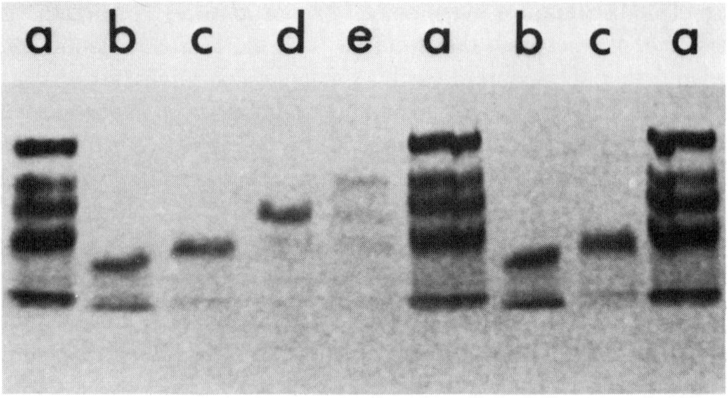

Figure 5. Excision of the chlorophyll *b*-containing PPCs from the Thornber and MARS fractionation systems and re-electrophoresis on the MARS system. Thylakoid membranes freshly treated with surfactant **(a)** were applied for comparison. The samples are CP-I **(b)** from the Thornber system and PPCs AB-3 **(c)**, AB-2 (**d**) and AB-1 (**e**) from the MARS system. This demonstrates that CP-II and AB-3, two similar chlorophyll *b*-containing PPCs, do not co-migrate and are presumably not identical. The AB-3 and AB-2 complexes also appear to be different, corroborating the previous suggestion (7) that they belong to a different oligomeric series of PPCs. Re-electrophoresis of the AB-1 complex (difficult to see) appears to give rise to complexes co-migrating with AB-1, AB-2 and AB-3, indicating that it is an oligomeric complex.

or three excised slices are dropped into the cuvette and arranged so that they sit in the bottom. A plastic tooth-pick or a stiff piece of small gauge plastic tubing is useful for arranging the slices and will not scratch the cuvette. The cuvette is positioned such that the spectrophotometer light beam passes through the gel slices and an absorption spectrum is recorded. It is convenient to have a cuvette which is black-masked and in which the bottom of the sample compartment is 9 or 10 mm above the bottom of the cuvette to ensure that the light beam passes through the gel slices. Since the pathlength of the sample is unknown and the concentration within the slice is not uniform, the absorption spectra are not absolute spectra and cannot be used to determine concentrations. These absorption spectra are useful in determining the pigment composition of the PPCs and providing a criterion for identification.

6.5 Polypeptide Analysis

A number of different systems are commonly used to analyse thylakoid membrane polypeptides. These systems are not equivalent and the results obtained may be dependent on the system used. The most widely used system is probably that developed by Chua (31) which usually gives excellent results. In recent studies on the apoproteins of photosystem I, the author in collaboration with Dr Philip Thornber's laboratory has developed a modification of the system of Laemmli (32) that appears to give improved resolution of these polypeptides (M.P.Skrdla, J.P.Markwell and J.P.Thornber, in preparation). This electrophoretic fractionation system is presented below. The best course in polypeptide analysis would be to use two different fractionation systems and compare the results.

Table 6. Media for the Electrophoretic Analysis of the Polypeptide Composition of Pigment-protein Complexes.

40% Acrylamide

1. Dissolve 80 g of acrylamide and 2.13 g of N,N′-methylenebisacrylamide in 120 ml of water and dilute to a final volume of 200 ml.

Lower Buffer (3 M Tris-Cl, 0.8% SDS, pH 8.8)

2. Dissolve 93.5 g of Tris base and 2 g of SDS in approximately 150 ml of water.
3. Adjust pH to 8.8 with concentrated HCl and dilute to a final volume of 250 ml.

Upper Buffer (1.0 M Tris-Cl, 0.8% SDS, pH 6.8)

4. Dissolve 12.12 g of Tris base and 0.8 g of SDS in approximately 75 ml of water.
5. Adjust pH to 6.8 with concentrated HCl, and dilute to a final volume of 100 ml.

10-fold Concentrated Reservoir Buffer (0.25 M Tris, 192 mM Glycine, 1% SDS, pH ~8.3)

6. Dissolve 60.6 g of Tris base, 288 g of glycine and 20 g of SDS in water and adjust to a final volume of 2 l. The resulting pH will be approximately 8.3 without adjustment. This solution is 10-fold concentrated: mix one volume with nine volumes of water prior to use.

Sample Denaturing Solution

7. Mix 0.25 ml of upper buffer, 0.6 ml of 20% SDS, 0.2 ml of 1 M dithiothreitol and 0.95 ml of a solution containing 50% glycerol and 0.1% bromophenol blue. Make fresh denaturing solution daily.

6.5.1 *Reagents and Procedures for Electrophoretic Analysis of PPC Polypeptide Composition (Modification of the System of Laemmli)*

Recipes are shown in *Table 6*.

6.5.2 *Gel Preparation*

(i) To prepare the lower gel, measure with water the volume necessary to fill the cavity between the gel plates up to a distance of 1 cm from the bottom of the well-forming comb.

(ii) Prepare the lower gel from two solutions of different urea and acrylamide concentrations using a small gradient-forming device and a peristaltic pump (6). The gradient of acrylamide (8 – 15%) is stabilised during polymerisation by the presence of the urea gradient (3 – 6 M).

(iii) Prepare the 'heavy' and 'light' lower gel solutions (*Table 7*) and incubate in a 30°C water bath with gentle swirling to dissolve the urea.

(iv) Place an amount of 'heavy' solution equal to half the total desired volume of lower gel in the chamber of the gradient-forming device closest to the outlet. Open the intervening stopcock and allow the space between the chambers to fill. Then close the stopcock.

(v) To initiate polymerisation, add 35 μl of 10% ammonium persulphate per 10 ml of gel solution. First add to the 'heavy' solution in the gradient maker and then add to an equal volume of 'light' solution and pour into the second chamber of the gradient former.

(vi) Mix the 'heavy' solution vibrationally or rotationally and turn on the peristaltic pump to a speed sufficient to move the solution into the gel apparatus in approximately 5 – 10 min (6).

(vii) Gently layer water onto the top surface of the lower gel, and allow the gel to polymerise for at least 1 h. At this point the gel can be used the same day, stored 1 or 2 days at room temperature, or stored for as long as 1 week in the cold.

Table 7. Gel Solutions for Electrophoretic Analysis of the Polypeptide Composition of Pigment-protein Complexes.

	Heavy solution	*Light solution*	*Upper gel*
Urea	11.5 g	5.8 g	7.7 g
Lower buffer	4.0 ml	4.0 ml	–
Upper buffer	–	–	2.0 ml
40% Acrylamide	12.0 ml	6.4 ml	1.6 ml
Water	7.2 ml	17.2 ml	6.3 ml
TEMED	8.5 μl	8.5 μl	16.25 μl
10% Ammonium persulphate	*	*	50 μl

* See text.

Cover gels that will be stored with plastic film ('cling film') to prevent the water on top of the gel from evaporating.

(viii) To pour the upper gel, remove the water from the top of the lower gel. Mix the upper gel solution (*Table 7*).

(ix) Incubate at 30°C with gentle swirling to dissolve the urea, add 50 μl of 10% ammonium persulphate and pour into the gel apparatus.

(x) Insert the sample well-forming comb and allow to polymerise for approximately 20 min before removing the comb.

(xi) Add reservoir buffer to the sample wells to prevent them from drying out.

(xii) Leave the gel at room temperature for another 30 – 45 min before removing the lower spacer and placing in the running assembly.

6.5.3 *Sample Denaturation*

(i) Mix equal volumes of denaturing solution and sample at a protein concentration of 2 mg ml^{-1} or less.

(ii) Incubate the samples at 60°C for 1 h following addition of denaturing solution.

(iii) Put slices of gel containing PPCs in a small tube and add sufficient denaturing solution to just cover the slice.

(iv) Incubate at room temperature for 1 h to allow the denaturing components to infiltrate the slice and then incubate for 1 h at 60°C.

(v) For very dilute samples, add sufficient trichloroacetic acid to bring the final concentration to 10%.

(vi) Incubate in the cold for 1 h and centrifuge.

(vii) Decant the supernatant taking care not to resuspend the pelleted protein, which is not always visible.

(viii) Invert the centrifuge tube, allow to drain and carefully wipe excess liquid from the inside upper-half of the tube.

(ix) Add a small amount of denaturing buffer and resuspend the protein by gentle shaking.

(x) The residual acid in the tube will turn the bromophenol blue in the denaturing solution yellow in colour. Add small volumes (2 – 5 μl) of 4 M NaOH to the solution until the colour is blue again.

(xi) Incubate for 1 h at 60°C before loading onto the gel.

6.5.4 *Electrophoresis*

Just prior to loading the samples onto the gel, use a small pipette and flush the sample wells with buffer from the reservoir. This is necessary because urea diffuses from the upper gel and the solution in the sample wells may become more dense than the samples. After loading, electrophoresis is carried out at a constant current. We find it convenient to load samples in the late afternoon and adjust the current such that the fractionation is complete (the bromophenol blue has migrated to the bottom of the gel) at mid-morning of the next day. The required current will depend on the length and thickness of the gel and must be determined empirically. Because of the low amounts of protein present in re-electrophoresed gel slices, it is necessary to visualise proteins by silver staining (6).

7. FACTORS AFFECTING RESOLUTION AND INTERPRETATION

7.1 **Free Chlorophyll**

It is believed that virtually all of the chlorophyll in the mature higher plant thylakoid membrane is complexed with protein to form PPCs. The free pigment observed following fractionation of PPCs by most fractionation systems results from disruption of the native complexes. It is currently unclear whether some of the chlorophyll is intercomplex in location, i.e., whether different PPCs form stable associations which include cooperative binding of pigment molecules. If this is the case, then the 10% free pigment observed with the MARS and Anderson systems may reflect real limits imposed by the molecular architecture. The diminished free pigment observed with the Deriphat 160 system may be the result of fractionating supramolecular PPCs in which the cooperatively complexed pigments are as yet relatively undisturbed.

In any case, the generation of free pigment is a fact of life with the most frequently used fractionation systems. This does, however, place some limitation on the ability to quantitate PPCs. If the amount of free pigment generated with a particular combination of plant material and fractionation system is appreciable (e.g. 20%) it is difficult to have faith in changes in the proportion of the various PPCs which may only amount to 5 or 10%. At present there is little that can be done to eliminate free chlorophyll during fractionation, so we must accept the limitations that this imposes on our ability to utilise this method for quantitative purposes.

7.2 **Ratio of Surfactant to Protein or Chlorophyll**

As can be seen from the fractionation procedures presented in this chapter, it is usual that the amount of detergent or surfactant to be added in order to solubilise thylakoid membranes is determined on the basis of chlorophyll present in the sample. The primary reason for this is that the assay for chlorophyll is rapid and convenient. However, it should be kept in mind that it would probably be more appropriate to use protein (or lipid) concentration as a basis for the amount of surfactant necessary for solubilisation. When working with any one plant system, the distinction may not be important, since the ratio of protein to chlorophyll in the thylakoid membranes will be relatively constant. However, this ratio can be variable when examining developing or greening systems, when comparing plants of different species, with plants grown under different conditions and photosynthetic mutants. Using these systems it might be appropriate to measure the protein concentration of the membrane and solubilise with constant ratios of surfactant to protein. This is not always practical. The Bradford (33) assay for protein is interfered with by the presence of chlorophyll which absorbs in the same region of the visible spectrum as the measuring dye. The Lowry (34) assay modified by the addition of SDS to the alkaline reagent (35) will accurately measure membrane protein concentrations, but the absorbance produced should be determined at 725 or 750 nm to avoid similar interference from chlorophyll. The major disadvantage of this latter assay is that it takes approximately 1 h and this amount of delay may adversely affect the following fractionation unless using a system in which the samples can be frozen and stored. These problems can often be overcome by empirically varying the surfactant to chlorophyll ratios, but the researcher should be aware of this problem and keep it in mind when comparing plant systems such as those described.

7.3 **Temperature Effects**

The Thornber, MARS and Deriphat 160 fractionation systems are only moderately sensitive to temperature as long as the solubilisation is carried out in the cold. The gels should be allowed to return to room temperature if the gel solution was incubated at super-ambient temperatures prior to polymerisation. The Anderson system, however, is very sensitive to the temperature of the gel during the electrophoresis. Gels run at room temperature rather than in the cold have much more free chlorophyll and much less pigment migrating with the PPCs; LHCP-3 appears to be particularly susceptible to this phenomenon. If this system is to be run regularly, it might be worthwhile investing in an electrophoretic apparatus that includes a cooling plate for water circulation and a refrigerated circulator to control rigidly the temperature during the fractionation.

7.4 **Environmental Effects**

Plants are sensitive to their environment and this can be reflected in their composition of PPCs and the ability of a particular fractionation system to produce repeatable results. Environmental factors normally considered include photoperiod, light quality and quantity, temperature, water availability, etc., and the reader is directed to a recent discussion of their effects on PPCs (36). The ability to control these factors will often result in more reproducible PPC fractionation.

One must also be aware of other environmental changes. While working on PPC fractionation with tobacco plants grown in a glasshouse, we observed that treatment of the plants with an insecticide to control the white fly population adversely affected the fractionation pattern. We have also observed the fractionation of additional PPCs apparently derived from photosystem I with several plant species that was apparently season-dependent. These additional complexes were observed during the month of June on several successive years. The basis of these seasonal effects is not understood and it is difficult to study something so transient and ephemeral. It is clear, however, that careful control of plant growth conditions is necessary for consistent PPC fractionation and that changes observed in plants collected in the field can be from a number of interrelated environmental and developmental factors.

8. ACKNOWLEDGEMENTS

Many of the fractionation systems described in this chapter were developed in the laboratory of J. Philip Thornber, University of California-Los Angeles with the assistance of Sally Reinman, Russell Boggs, Donald Miles, Merri Skrdla and Saeid Nourizadeh. This research has been funded over a period of years by grants awarded to Philip Thornber and the author from the National Science Foundation and the Competitive Research Grants Office of the U.S. Department of Agriculture.

9. REFERENCES

1. Rabinowitch,E.I. (1945) *Photosynthesis and Related Processes,* Vol. **1**, published by Interscience Publishers, New York, p. 355.
2. Hiller,R.G. and Goodchild,D.J. (1981) in *The Biochemistry of Plants*, Vol. **8**, Hatch,M.D. and Boardman,N.K. (eds.), Academic Press, London and New York, p. 1.
3. Thornber,J.P. (1975) *Annu. Rev. Plant Physiol.*, **26**, 127.
4. Thornber,J.P., Markwell,J.P. and Reinman,S. (1979) *Photochem. Photobiol.*, **29**, 1205.
5. Markwell,J.P., Thornber,J.P. and Boggs,R.T. (1979) *Proc. Natl. Acad. Sci. USA*, **76**, 1233.
6. Hames,B.D. (1983) in *Gel Electrophoresis of Proteins – A Practical Approach*, Hames,D.B. and Rickwood,D. (eds.), IRL Press, Oxford and Washington, D.C., p. 1.
7. Markwell,J.P., Nakatani,H.Y., Barber,J. and Thornber,J.P. (1980) *FEBS Lett*, **122**, 149.
8. Rabinowitch,E.I. (1951) *Photosynthesis and Related Processes*, Vol. **2**, published by Interscience Publishers, New York, p. 623.
9. Bartzatt,R.A., Yang,C.M. and Markwell,J.P. (1983) *Biochim. Biophys. Acta*, **725**, 341.
10. Anderson,J.M, Waldron,J.C. and Thorne,S.W. (1978) *FEBS Lett.*, **92**, 227.
11. Miles,C.D., Markwell,J.P. and Thornber,J.P. (1979) *Plant Physiol.*, **64**, 690.
12. Reinman,S. and Thornber,J.P. (1979) *Biochim. Biophys. Acta*, **547**, 188.
13. Thornber,J.P. and Highkin,H.R. (1974) *Eur. J. Biochem.*, **41**, 109.
14. Hopkins,W.G., Hayden,D.B. and Neuffer,M.G. (1980) *Z. Pflanzenphysiol.*, **99**, 417.
15. Metz,J.G., Kreuger,R.W. and Miles,D. (1984) *Plant Physiol.*, **75**, 238.
16. Hiller,R.G., Pilger,T.B.G. and Genge,S. (1978) in *Chloroplast Development*, Akoyunoglou,G. and Argyroudi-Akoyunoglou,J.H. (eds.), Elsevier, Amsterdam, p. 215.
17. Argyroudi-Akoyunoglou,J.H. and Castorinis,A. (1980) *Arch. Biochem. Biophys.*, **200**, 326.
18. Arntzen,C.J., Armond,P.A., Briantais,J.M. and Novitzsky,W.P. (1977) *Brookhaven Symp. Biol.*, **28**, 316.
19. Leong,T.Y. and Anderson,J.M. (1983) *Biochim. Biophys. Acta*, **723**, 391.
20. Markwell,J.P., Reinman,S. and Thornber,J.P. (1978) *Arch. Biochem. Biophys.*, **190**, 136.
21. Arnon,D.I. (1949) *Plant Physiol.*, **24**, 1.
22. Thornber,J.P. (1970) *Biochemistry (Wash.)*, **9**, 2688.
23. Thornber,J.P., Gregory,R.P.F., Smith,C.A. and Bailey,J.L. (1967) *Biochemistry (Wash.)*, **6**, 391.
24. Hiller,R.G., Genge,S. and Pilger,D. (1974) *Plant Sci. Lett.*, **2**, 239.
25. Neville,D.M. (1971) *J. Biol. Chem.*, **246**, 6328.
26. Anderson,J.M. (1980) *Biochim. Biophys. Acta*, **591**, 113.
27. Anderson,J.M., Brown,J.S., Lam,E. and Malkin,R. (1983) *Photochem. Photobiol.*, **38**, 205.
28. Delepelaire,P. and Chua,N.H. (1979) *Proc. Natl. Acad. Sci. USA*, **76**, 111.
29. Siefermann-Harms,D. and Ninnemann,H. (1979) *FEBS Lett.*, **104**, 72.
30. Bennett,J., Markwell,J.P., Skrdla,M.P. and Thornber,J.P. (1981) *FEBS Lett.*, **131**, 325.
31. Chua,N.H. (1980) in *Methods in Enzymology*, Vol. **69**, San Pietro,A. (ed.), Academic Press Inc., London and New York, p. 434.
32. Laemmli,U.K. (1970) *Nature*, **277**, 680.
33. Bradford,M. (1976) *Anal. Biochem.*, **72**, 248.
34. Lowry,O., Rosebrough,N., Farr,A. and Randall,R. (1951) *J. Biol. Chem.*, **193**, 265.
35. Dulley,J.R. and Grieve,P.A. (1975) *Anal. Biochem.*, **64**, 136.
36. Baker,N.R. and Markwell,J.P. (1985) in *Topics in Photosynthesis, Vol. 6, Photosynthetic Mechanisms and the Environment*, Barber,J. (ed.), Elsevier Press, Amsterdam, p. 49.

CHAPTER 4

Spectroscopy

M.F. HIPKINS and N.R. BAKER

1. INTRODUCTION

Spectroscopy is the name given to a family of analytical techniques that has a common basis: the interaction of electromagnetic radiation with matter. The different types of spectroscopy are defined principally by the range of wavelengths that is used; from the X-ray region (0.1 – 100 nm) to microwaves (0.1 mm – 10 cm). The energy of the electromagnetic radiation is proportional to the inverse of the wavelength, therefore these types of spectroscopy correspond to probing the system under study with different amounts of energy. Thus X-ray spectroscopy is used to study processes which involve large amounts of energy, for example the transitions of inner shell electrons of atoms. Conversely microwave spectroscopy is used to investigate molecular rotations, where the energy changes involved are rather small. Spectroscopy in the visible region of the electromagnetic spectrum (in the approximate wavelength range 380 – 750 nm) is used as an analytical tool for transitions between outer shell electrons in atoms and electronic orbitals in molecules. The latter processes lie at the very heart of photosynthesis – as a result visible absorption spectroscopy is widely used as a non-destructive technique in photosynthesis work.

In addition to absorption spectroscopy, fluorescence spectroscopy is used in photosynthesis research. This is because photosynthesis involves the capture of light by complex organic pigments some of which re-emit a small proportion of the light they absorb in a process called fluorescence. Fluorescence spectroscopy, a technique closely allied to absorption spectroscopy, exploits this phenomenon, using in particular the fluorescence from chlorophyll *a*. Fluorescence is a wasteful process for the plant, but a powerful non-destructive analytical technique since it can give information on photophysical processes, photosynthetic electron transfer and even the fixation of carbon dioxide.

More specialised spectroscopic techniques are also used, and these include absorption spectroscopy using polarised light (linear and circular dichroism), electron paramagnetic resonance spectroscopy (e.p.r., used in the study of chemical species containing unpaired electrons) and nuclear magnetic resonance spectroscopy (n.m.r., used to study compounds containing suitable elements, typically H, C and P in biological systems). In this chapter we shall describe in some detail both absorption and fluorescence spectroscopy since they have wide application in photosynthesis. The equipment necessary for some of these techniques is normally available in well-found laboratories. In addition we give some design criteria for the construction of the less sophisticated items. The other more specialised techniques will be described only briefly. In contrast to the other chapters in this book, we shall emphasise the theory of absorption and fluorescence spectroscopy, because the techniques themselves, and the analysis of the data they provide, depend on a proper understanding of the underlying physical principles.

2. THE INTERACTION OF ELECTROMAGNETIC RADIATION WITH MATTER

Light or, more correctly, electromagnetic radiation, can be thought of in two ways: either as waves or as particles. When light interacts with light then the description in terms of wave theory works well. The wave theory describes light as coupled sinusoidal oscillations of an electric field and a magnetic field; the electric and magnetic fields being at right-angles to each other, and also to the direction of propagation of the light beam. In contrast, when light interacts with matter and transfers energy to it, then it is necessary to use the 'quantum description' and to think about light as a stream of energy-carrying particles. Each of these particles is called a photon. It has no mass, but carries a discrete well-defined packet (quantum) of energy. The wave theory and the quantum theory are closely related, since the energy of a quantum is given by

$$E = h\nu = hc/\lambda \qquad \text{Equation 1}$$

where ν is the frequency of the radiation (sec^{-1}), c is the velocity of light (2.988×10^8 m sec^{-1}), λ is the wavelength (m) and h is Planck's constant (6.626×10^{-34} J sec). The electromagnetic spectrum comprises electromagnetic radiation whose wavelengths vary from 10^{-15} m (highly energetic cosmic rays) to 10^3 m (radio waves). Only a small section of the electromagnetic spectrum can be detected by the human eye, in the approximate wavelength region 380 – 720 nm; this band of wavelengths is called the visible region of the spectrum. The preferred wavelength unit in the visible region is the nanometre (nm, 10^{-9} m), which in older textbooks and articles is sometimes written as mμ (millimicron). The Angstrom (Å = 0.1 nm) is no longer used. A photon of light in the middle of the visible region (λ = 550 nm) has an energy of 3.62×10^{-19} J, and a mole of such photons (6.023×10^{23} photons; formerly called an Einstein) has an energy of 2.18×10^5 J. Photosynthetically active radiation (PAR) is generally taken to comprise wavelengths in the range 400 – 700 nm, although wavelengths in the 350 – 400 nm range can drive a small amount of carbon dioxide assimilation. The favoured unit for PAR is the quantum flux density (quantum fluence rate) measured in μmol photons m^{-2} sec^{-1} (1; Appendix I).

2.1 **Absorption of Radiation by Atoms and Molecules**

2.1.1 *Atoms*

When a substance absorbs radiation its energy content increases. Conversely, when it emits radiation its energy content decreases. In the quantum theory, it is found that the energy content of atoms and molecules cannot assume any arbitrary value, but must be confined to certain fixed values, called energy levels. The energy levels are characteristic of the particular atom or molecule under study. In the classical model of the atom, energy levels correspond to the orbits of increasing radius that electrons can describe round the nucleus. When energy is absorbed it is used to promote electrons from lower energy levels to higher energy levels (*Figure 1*). The lowest energy

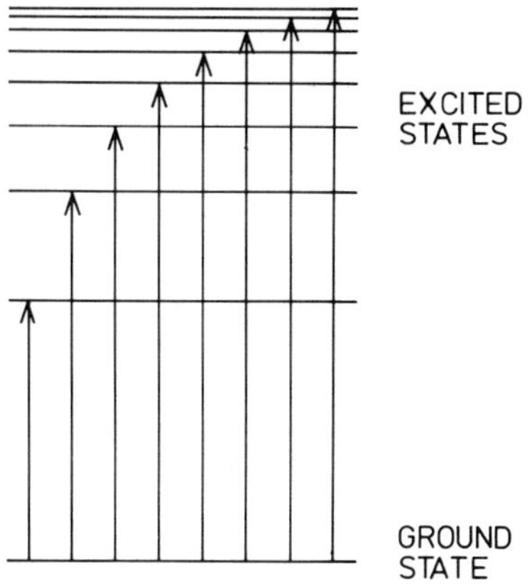

Figure 1. Schematic diagram showing the electronic energy levels of an atom. The upward arrows represent the excitation of an electron from the ground state to one of the excited states following the absorption of a photon of light whose energy exactly matches the energy gap. Light emission (fluorescence) takes place when radiative de-excitation of the excited electron to the ground state occurs.

level is called the ground state. When energy is absorbed, and an electron undergoes a transition from a lower energy level to a higher one, the atom becomes 'excited'. The wavelengths of light that an atom can absorb are defined by the exact energy gap between the ground and excited states:

$$E_e - E_g = hc/\lambda \qquad \text{Equation 2}$$

where E_e and E_g are the energies of the excited state and ground state, respectively (*Figure 1*). Clearly, the atom or molecule has then taken a part of the energy of a beam of light and converted it to electronic excitation energy. The intensity of the beam is therefore diminished by the amount the atom or molecule has absorbed. In general an atom can undergo a number of such transitions from the ground state to excited states of various energies, and it can therefore absorb several wavelengths of light, leading to an absorption spectrum. The relative intensity of absorption lines or bands depends on the relative probability of a transition taking place between two energy levels. A detailed discussion of this topic will not be given here, but has been covered well for the biologist by Clayton (2).

Once in an excited state, the electron tends to return to the stable, ground state. In atoms this can be by radiationless decay (heat) to the surroundings, or can be accompanied by re-emission of light. In the latter case the energy of the emitted photon is exactly equal to the energy difference between excited state and ground state (*Figure 1*). In the simplest case, that of the hydrogen atom, a number of groups of emission lines is seen, which correspond not only to transitions between excited states and the ground state but also to transitions between higher and lower excited states. The emission lines taken together constitute the emission spectrum of hydrogen.

2.1.2 *Molecules*

The absorption spectra of atoms that contain more than one electron are more complex than that of hydrogen, but still consist of numbers of absorption lines. The absorption processes in molecules are analogous to those of atoms, but the resulting spectra are more complex again. Electrons in molecules are found in molecular orbitals, which correspond to the orbits of atoms. Some electrons are involved in covalent bonding, and are in so-called σ orbitals, which correspond to saturated bonds between carbon atoms. On excitation, an electronic transition to an excited, anti-bonding σ^* orbital can take place. Normally, the σ orbitals are fully occupied in the ground state, whilst the σ^* orbitals are completely vacant. The $\sigma \rightarrow \sigma^*$ transitions are normally seen in the u.v. region of the spectrum. Other electrons are involved in π orbitals, particularly in compounds with conjugated double bonds. As a result of hybridisation, the bonding π orbitals tend to be delocalised over the part of the molecule with the conjugated system. As in the case of σ orbitals, π orbitals are normally fully occupied in the ground state, but π^* orbitals are fully vacant. Excitation by light can lead to a $\pi \rightarrow \pi^*$ transition. There are also n orbitals, which are associated with the lone-pair electrons of oxygen and nitrogen. Both $\pi \rightarrow \pi^*$ and $n \rightarrow \pi^*$ transitions involve absorption of light in the u.v. and visible regions of the spectrum.

The molecular absorption spectra of pigments of biological interest generally consist of quite wide bands of absorption rather than the sharp absorption lines of atoms. These bands arise from two components of the molecular energy (apart from electronic energy) which do not have counterparts in atoms. The first component is due to the vibrations of atomic nuclei within molecules about their 'mean' positions; the second component is due to the rotations of molecules about their centres of gravity. The energy spacing between vibrational levels is up to an order of magnitude smaller than that of electronic energy levels; rotational energy spacings are about an order of magnitude smaller again. Whilst all these levels are discrete, the result is that each electronic energy level is subdivided by two perturbations, leading to a large number of rather closely spaced levels which effectively make an absorption band. These points are illustrated in the hypothetical molecular energy level diagram (*Figure 2*). Using i.r. spectroscopy it is possible to distinguish molecular vibrations and rotations in gases, but for large molecules

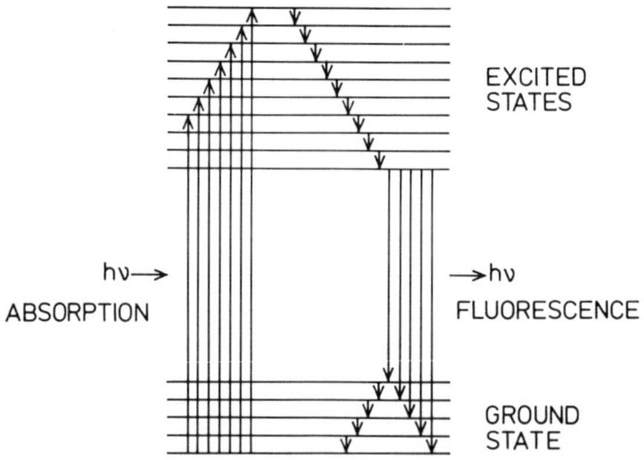

Figure 2. Schematic energy-level diagram for a molecule. Both ground and excited states have many closely-spaced substates due to molecular vibration and rotation. Excitation of an electron occurs from the lowest substate of the ground state to any one of the substates of the excited states: excitation is followed by rapid radiationless de-excitation to the lowest substate of the excited state. Radiative decay (fluorescence) occurs when de-excitation to one of the substates of the ground state takes place; this is then followed by further radiationless decay to the lowest substate of the ground state.

in solution the moment of inertia of the molecule, and collisions with solvent molecules, tend to rule out the observation of fine-structure in the absorption spectrum.

In molecules two types of excited states exist: the short-lived singlet state where pairs of electrons in the molecular orbital have opposite spin, and triplet states where the spins of the electrons are aligned. Excited triplet states typically have much longer lifetimes than singlet states, and have lower energy. Conversion between singlet and triplet states is called 'inter-system crossing' (see Section 2.3).

2.1.3 *Absorption by the Photosynthetic Apparatus*

The major photosynthetic pigments are chlorophylls and carotenoids, with the bile pigments occurring in red algae and blue-green algae (cyanobacteria). The light-absorbing sections (chromophores) of these molecules possess extensive conjugated double-bond systems. The difference in energy between the π and π^* orbitals depends on the extent of the conjugated system as judged by the number of continuous double bonds; the more double bonds there are, the smaller the energy gap between ground and excited states, and hence the longer the wavelength of light which will excite a transition. The phenomenon of increasing wavelength of absorption with the increasing extent of double bonds is known as the 'bathochromic' shift.

Whilst the most significant contribution to the absorption of a photosynthetic organism is made by chlorophylls, carotenoids and (where present) phycobilins, small but important contributions are made by transition metal complexes, like cytochromes (complexes of protein and haem iron) and plastocyanin (a copper-containing protein) both of which are involved in photosynthetic electron transport. The absorption properties of these complexes are especially interesting, since they depend on the oxidation-reduction conditions of the metal ion.

2.2 **Decay of Excited States**

Once a molecule is in a metastable excited state there are several ways in which the excitation energy may be lost, each of which leads to the attainment of a stable, lower energy state. The fate of the electron in an excited state will be dictated by the relative rate constants for each of the de-excitation processes, those with the higher rate constants being more probable. When excited, the electron will initially be in one of the higher rotational or vibrational sub-states of the excited state (*Figure 2*). The rate constant for radiationless decay from one rotational or vibrational sub-state to the next lower one is very large ($\sim 10^{15}$ sec^{-1}) which means that excited electrons very quickly lose energy in a number of small steps, until they come to the lowest sub-state of the first excited singlet state (*Figure 2*). If the molecule does not interact with other molecules, then from this lowest sub-state the excited electron has three main routes that it can take: firstly, it can undergo a transition accompanied by the emission of a photon from the excited state to the ground state, in a process called fluorescence. Secondly, there can be intersystem crossing to the triplet manifold. Thirdly, the electron can undergo non-radiative decay to the ground state. In addition, if the molecule is a part of the photosynthetic system, or any other photobiological complex, the excitation energy can either be transferred to a neighbouring molecule, or alternatively be used in a photochemical reaction.

2.3 Fluorescence

As noted in the previous section, fluorescence is one of the ways in which an excited molecule can regain the ground state. It involves the emission of radiation, which can be detected fairly readily. It should be realised that fluorescence is only one of a number of competing de-excitation processes, so that the information which fluorescence gives reflects all of the de-excitation processes.

Some molecules, like the majority of carotenoids, do not give significant fluorescence, whilst others, like chlorophyll, fluoresce strongly. The tendency to lose excitation energy as fluorescence is quantified by the fluorescence yield, ϕ_f, so that

$$F = \phi_f I \qquad \text{Equation 3}$$

where I is the amount of light absorbed, and F is the fluorescence intensity. It is relatively difficult to measure absolute values of ϕ_f, and often it is unnecessary. The most normal method is to compare the fluorescence of an unknown sample with that of a standard whose fluorescence yield is known, in an identical experimental system.

The value of fluorescence yield will lie between 0 and 1; its value will be determined by the nature of the other processes which compete for de-excitation of the excited state. Writing k_f for the rate constant for fluorescence, then for a simple isolated molecule

$$\phi_f = k_f/(k_f + k_h + k_{isc}) \qquad \text{Equation 4}$$

where k_h is the rate constant for radiationless decay to heat, and k_{isc} is the rate constant for intersystem crossing from the excited singlet state to the triplet state.

This expression for fluorescence has little applicability to fluorescent chlorophyll molecules in a photosynthetic organism, but it does illustrate the important point that the fluorescence yield depends not only on the value of k_f, but on the values of the rate constants for all the other processes involved in the de-excitation of the excited state. For example, if k_{isc} were to increase, then ϕ_f would decrease due to this factor, not due to a change in k_f itself.

The fluorescence yield can also be deduced from the lifetime of the decay of the fluorescence after a brief exciting flash. To understand how the relationship between yield and lifetime can be made, it is necessary to imagine a hypothetical situation: suppose a pigment existed where the only route for the decay of the excited state were *via* fluorescence. The lifetime of the excited state, τ_o, would then be given by:

$$\tau_o = 1/k_f \qquad \text{Equation 5}$$

where τ_o is called the 'natural' lifetime of the excited state. In reality, fluorescence is only one of a number of competing decay routes for the excitation, so that in general the real lifetime, τ, is given by:

$$\tau = 1/(k_f + k_h + k_{isc}) \qquad \text{Equation 6}$$

Because more processes compete for the decay of excited states, the denominator of the equation is larger, so that $\tau < \tau_o$. The fluorescence yield is then:

$$\phi_f = k_f/(k_f + k_h + k_{isc}) = \tau/\tau_o \qquad \text{Equation 7}$$

For chlorophyll in organic solution, τ_o has been estimated to be about 15 nsec, whilst τ for chlorophyll *in vivo* is of the order of 1 nsec (Section 7.3.3), giving ϕ_f of around 0.05.

In oxygen-evolving organisms it is found that at room temperature the fluorescence emission comes from the chlorophyll molecules in the pigment-protein complexes that are associated with photosystem II, whilst photosystem I is essentially non-fluorescent under normal circumstances. Thus in considering the factors which tend to alter fluorescence yield in plants, photosystem II is central.

Excited chlorophyll molecules in photosystem II can lose energy in four main ways:

(i) by fluorescence (rate constant k_f);
(ii) by radiationless decay to heat (k_h);
(iii) by energy transfer to the non-fluorescent photosystem I [$k_{t(II \rightarrow I)}$] and
(iv) by use in the photochemical reactions of photosystem II (k_p).

This last rate constant, k_p, is normally multiplied by P, the fraction of photosystem II reaction centres that is able to accept excitation energy. If all the reaction centres can accept energy, P = 1, whereas if none can, P = 0.

The fluorescence yield is given by

$$\phi_f = k_f/[k_f + k_h + k_{t(II \rightarrow I)} + k_p P] \qquad \text{Equation 8}$$

As noted above, alteration of any of the terms in the denominator of this expression will alter the value of ϕ_f. The most significant factor is often P, but $k_{t(II \rightarrow I)}$ also has a role to play in some circumstances.

For photosynthetic systems, which in higher plants consists of three major pigment beds [photosystem I, photosystem II and the light-harvesting chlorophyll protein complex associated with PSII (LHCII)], the expression relating fluorescence intensity (normally the experimentally measured parameter) and illumination intensity (Equation 3) needs modification to take into account the fraction of the incoming light energy that is directed to photosystem II. Thus

$$F = \beta \, \phi_f \, I \qquad \text{Equation 9}$$

where β is the fraction of light incident on the photosynthetic apparatus that is intercepted by, or directly transferred to, photosystem II.

2.3.1 *Fluorescence Excitation and Emission Spectra*

A fluorescence excitation spectrum measures the relative efficiency with which different wavelengths of light elicit fluorescence from a pigment. Normally, for a single pigment in solution, the absorption spectrum of the pigment corresponds to the excitation spectrum for fluorescence. In photosynthetic organisms, however, the accessory pigments (e.g. carotenoids and chlorophyll *b* in higher plants) capture light energy and direct it to the fluorescent chlorophyll *a* molecules of photosystem II. It should be no surprise, therefore, that the excitation spectrum of *in vivo* chlorophyll fluorescence corresponds closely to the action spectrum of photosystem II.

The fluorescence emission spectrum of chlorophyll shows two interesting points, both concerned with de-activation processes immediately after excitation (Section 2.2). *Figure 2* shows that after excitation, the first decay process is the very rapid radiationless decay from the energy level at which the excited electron arrives to the lowest vibrational sub-state of the first excited state. Since the sub-states of the first and higher excited states overlap, this means that irrespective of whether the excitation is to a higher (short wavelength) excited state, the excited electron loses energy with the consequence that fluorescence always occurs from the lowest (long wavelength) excited state. Therefore, although chlorophyll absorbs strongly in the blue and red regions of the spectrum, fluorescence is confined to the red region. Moreover, careful study of *Figure 2* will show that the average energy gap for excitation to the first excited state is greater than the energy gap for fluorescence. This means that the peak of the fluorescence emission is at a longer wavelength than the peak of absorption. The wavelength difference is called the Stokes' shift.

3. MEASUREMENT OF ABSORPTION

This section and the next (Section 4) consider the theory and practice of using the commercially available spectrophotometers found in most laboratories in the investigation of photosynthetic systems. The applications are limited to measurements in the steady-state, or to changes that are so slow they may be measured discontinuously. For rapidly varying absorption measurements, much more specialised equipment is required — the apparatus and techniques are considered in Section 5.

3.1 **The Beer-Lambert Law**

The Beer-Lambert law relates the absorbance of an optically clear (non-scattering) solution to its concentration. It is built on two simple postulates. Firstly, the Beer Law states that for a solution which has monochromatic light passed through it, for an infinitely thin slice of the solution the proportion of the incident light absorbed ($-dI/I$) is a constant, and is proportional to the concentration (c) of the sample. Secondly, the Lambert Law states that, in the situation described above, the fractional absorption is proportional to the thickness (dl) of the sample. Thus,

$$-dI/I = K_\lambda \, c \, dl \qquad \text{Equation 10}$$

where K_λ is a constant of proportionality that depends on the wavelength. On integration, using suitable limits

$$\ln [I_o/I_f] = K_\lambda cl \qquad \text{Equation 11}$$

where I_o is the intensity of light entering the sample, and I_f is the intensity leaving the sample. Remembering that $\ln y = 2.303 \log_{10} y$

$$\log [I_o/I_f] = (K/2.303)\, cl \qquad \text{Equation 12}$$

Writing the absorbance, A, for $\log [I_o/I_f]$, and putting $K/2.303 = \epsilon$, the extinction coefficient, then:

$$A = \epsilon \, cl \qquad \text{Equation 13}$$

Note that A is dimensionless, so that ϵ has dimensions of concentration^{-1} length^{-1}. ϵ varies with wavelength, being higher in regions of the spectrum where strong absorption occurs. The units of ϵ can take several forms, the most usual being mM^{-1} cm^{-1} or $M^{-1}\,cm^{-1}$. In older texts $\epsilon^{1\%}_{cm}$ may be found, signifying the extinction given by a 1% solution in a 1 cm cell. Conversion to a molar extinction coefficient can be made if the molecular weight of the substance is known. Absorbance is also known as optical density (OD), although at present absorbance is the preferred term. Of course, absorbance and the percentage transmission (T) of a solution are related, through:

$$A = -\log(T/100) \qquad \text{Equation 14}$$

Deviations from the Beer-Lambert law must be expected, particularly where the absorbance A is above 1.5. A good rule-of-thumb is not to work with absorbances significantly over 1.

3.2 Turbid Solutions

The above derivation of the Beer-Lambert law makes a number of assumptions: perhaps the most important of these is that the solution under consideration is 'optically clear', that is, it does not contain any substances or particles that are likely to scatter the measuring beam out of the optical path in the spectrophotometer. Solutions such as photosynthetically active chloroplast and membrane preparations which do scatter are called turbid. The effects of turbidity are 2-fold: firstly, light can be scattered out of the optical path so that it is not sensed by the detector, thus giving an apparent absorbance. Secondly, multiple scattering may occur so that even though no light is scattered right out of the measuring beam, the path length is significantly increased as the beam crosses the cell. Measurement of absorption in turbid samples will be considered below (Section 4.2).

3.3 Typical Spectrophotometers

All spectrophotometers, irrespective of their apparent degree of complexity, have a number of common components. These have the following functions:

(i) to generate monochromatic light of the desired wavelength,
(ii) to pass this through a cell containing the sample solution or a reference solution,
(iii) to measure the amount of light absorbed by the sample as compared with the reference, and
(iv) to compute the absorbance.

It should be noted that there is a distinct difference between a spectrophotometer and a colorimeter. A spectrophotometer uses essentially monochromatic light to measure the absorption, whilst a colorimeter uses a broad band of wavelengths selected by a simple colour filter. Since the Beer-Lambert law is only valid for monochromatic light, a colorimeter does not measure absorption, and should not be used for such measurements.

In most spectrophotometers quasi-monochromatic light is generated by a monochromator equipped with a diffraction grating which disperses polychromatic 'white' light from a source into its component wavelengths. The dispersing component is in some cases a prism. The light source is typically a deuterium lamp for work in the u.v. region of the spectrum ($\sim 190-350$ nm), with tungsten-halogen lamps used in the visible and near i.r. ($350-900$ nm). The monochromatic light passes through the cell and is detected by a photomultiplier tube, which has very high sensitivity. The output signal from the photomultiplier is then amplified with a circuit that at the same time takes the logarithm of the signal.

Commercially available spectrophotometers are of two main types: those simple single-beam spectrophotometers that measure absorption at a single wavelength, and more complex double-beam scanning spectrophotometers.

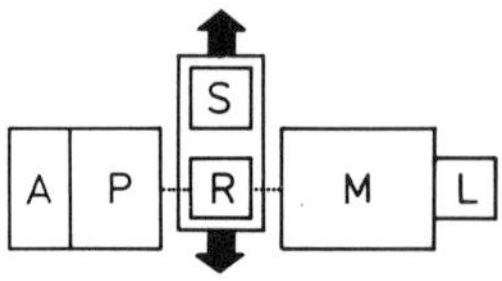

Figure 3. Schematic diagram of a simple single-beam spectrophotometer. L, lamp; M, monochromator; S, sample cell; R, reference cell; P, photodetector; A, amplifier. The cell carrier can normally slide back and forth (solid arrows) so that either the sample cell or the reference cell is in the light beam (dotted line).

3.3.1 *Single-beam Spectrophotometers*

A schematic diagram of a simple, single-beam spectrophotometer is shown in *Figure 3*, where the components described above are seen. In using such a simple instrument it is necessary to adjust the wavelength manually, and also to establish the value of I_0 (Equation 12) by determining the amount of light absorbed by the reference cell, which contains only the solvent. Typically, the procedure would involve the following steps.

(i) Select the desired wavelength.
(ii) Place the reference cell in the light path.
(iii) Using a shutter, cut off the measuring beam and set 0% transmission (infinite *A*).
(iv) Open the shutter, and using a slitwidth, high voltage or light intensity control, set 100% T (zero *A*).
(v) Close the shutter, remove the reference cell, replace it with the sample cell, open the shutter and read off the absorbance.

Some older instruments work on a 'null principle' where the absorbance value is adjusted until a meter reads zero. An absorption spectrum can be built up, albeit rather laboriously, by repeating the above procedure at each wavelength.

3.3.2 *Double-beam Scanning Spectrophotometer*

A more sophisticated instrument is the double-beam scanning spectrophotometer which can record an absorption spectrum automatically. The optical components are shown in *Figure 4*. The spectrophotometer has positions both for the reference cell (solvent only) and the sample cell (solvent plus sample). At each wavelength the measuring beam is optically switched so that it passes first through the reference cell, then through the sample cell in quick succession. The instrument determines the value of I_0 from the reference cell and compares it with I_f from the sample cell (Equation 12). The value of *A* is computed and passed to a chart recorder device which plots the value. When measurement at one wavelength is complete, the monochromator is advanced to the next wavelength by a motor, which also advances the chart recorder. The absorbance at the new wavelength is computed and plotted, and in this way an absorption spectrum is built up.

Modern scanning spectrophotometers have sophistications provided by microprocessors. For example, the cells used for the reference and sample solutions may not be completely matched (see Section 3.4.1), and this would give rise to artefactual absorption signals in certain regions of the spectrum. With some instruments it is poss-

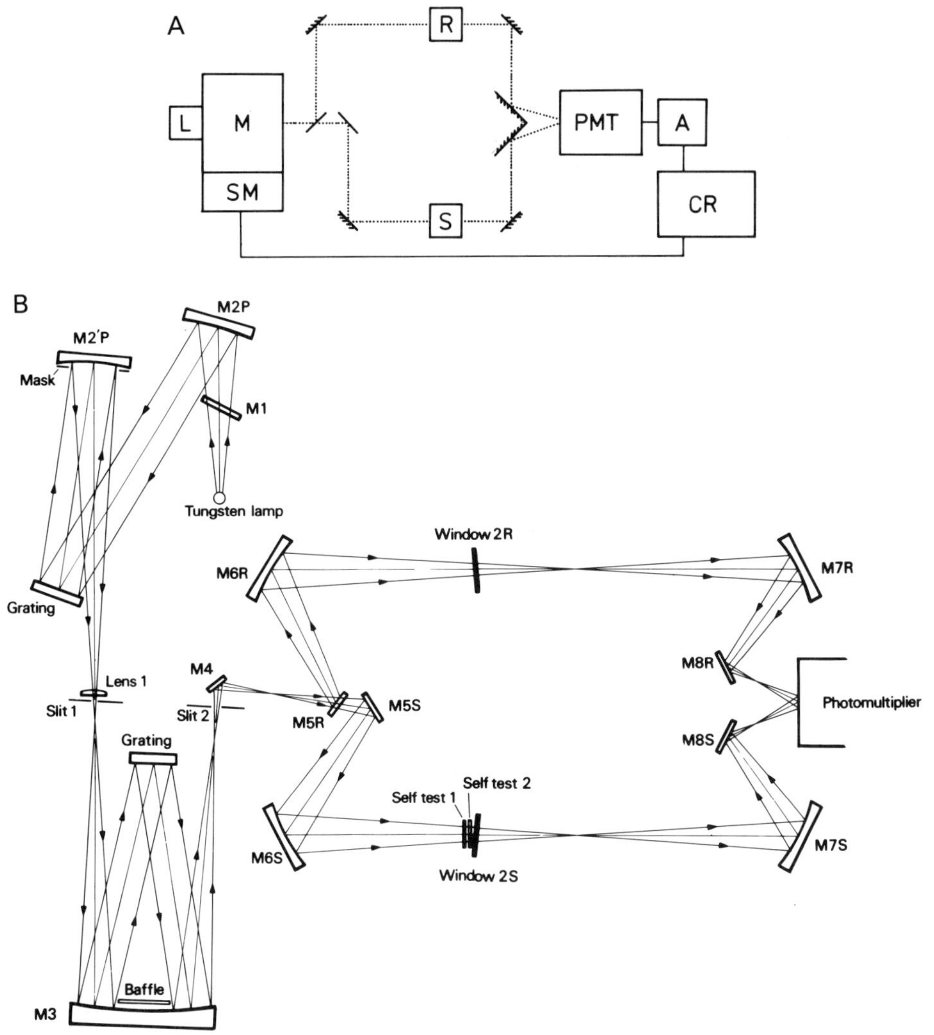

Figure 4. (**A**). Schematic diagram of a double-beam scanning spectrophotometer. L, lamp; M, monochromator; SM, stepper motor; R, reference cell; S, sample cell; PMT, photomultiplier tube; A, amplifier; CR, chart recorder. The light beam (dotted line) is optically switched so that it passes alternately through the reference and sample cells. The amplifier computes the absorbance at each wavelength; the stepper motor advances the monochromator to the next wavelength and at the same time moves the chart on the chart recorder. (**B**) The optical layout of a typical double-beam scanning spectrophotometer, the Pye-Unicam PU8800. The reference and sample cells are placed at the focal points between window 2 and mirror M7. Mirrors MR5 and MS5 serve to direct the measuring beam alternately through the reference and sample cells. (Diagram kindly supplied by Pye-Unicam Ltd., Cambridge).

ible to allow for this by measuring and storing a 'baseline' spectrum by having solvent in both the reference and sample cells. This 'baseline' can then be subtracted from subsequent absorption spectra which, provided the same cells are used, will thus be corrected for the differences between the cells. Similarly, it would be possible to record reference and sample spectra separately and to produce the real absorption spectrum by subtraction: this technique has sometimes been used when recording the absorption spectra of highly-scattering samples in laboratory-built spectrophotometers (Section 4.2.2).

3.4 General Practical Points on Making Spectrophotometric Measurements

3.4.1 *Spectrophotometer Cells*

Cells, or cuvettes, are made in several materials. The cheapest are plastic, but are not to be recommended since they have rather poor optical qualities. Moreover, they can be attacked by organic solvents, particularly acetone. The choice really lies between the different types of glass cells, and the main criterion is the wavelength range to be used. Standard glass becomes opaque below about 350 nm, so for measurements below this wavelength cells made from special silica glasses, which transmit down to 170 nm, should be used. At the long-wavelength end, no problems of opacity occur until about 2500 nm, although special materials for use in the i.r. are available. The main drawback of the special glass cells for use in the u.v. is their expense, particularly for fluorescence work where four clear sides are required.

For accurate absorption work, cells should be matched. This is normally a service offered by the manufacturer, where cells of similar optical characteristics are put in pairs to minimise absorption artefacts arising from cells with slightly different absorption properties.

The cleanliness of cells is also important. They should be rinsed and dried carefully after use, and fingermarks on the optical surfaces should be avoided. Cells should not be overfilled. Firstly, this can lead to drops of samples running down the side of the cell and causing artefacts, and secondly, the measuring light beam is relatively small and normally found at about the centre of the cell, so filling the cell to the brim is unnecessary. Where the volume of sample is less than about 3 ml (the volume required to fill a standard 10 mm pathlength cell), semi-micro cells can be used, with a volume of about 1 ml.

The samples should also be at room temperature. If they are cold, two artefacts may occur. Firstly, condensation may form on the windows of the cell, leading to an increase in scattering. Secondly, as the solution warms up, dissolved gasses may come out of solution and form bubbles which also tend to scatter light out of the measuring beam.

3.4.2 *Sample Solutions*

Sample solutions should be optically clear, and not contain any suspended particulate matter. Thus, solutions should be filtered or centrifuged to remove such particles. The problems of turbid solutions have been mentioned in Section 3.2.

4. APPLICATIONS OF ABSORPTION SPECTROSCOPY IN PHOTOSYNTHESIS

4.1 Estimation of Chlorophyll

In the majority of biochemical measurements on photosynthesis it is useful to express rates of activity per unit chlorophyll. Determination of the chlorophyll content of a sample is therefore one of the most important routine measurements. A determination involves extracting the material with a suitable organic solvent (chlorophyll is insoluble in water), measuring its absorption, and then using the extinction coefficients for the component pigments to calculate the concentration of chlorophyll. There are a number of sources of error which need to be taken into account (3).

In general an organic solvent will extract not only chlorophylls but other pigments as well. In green algae and higher plants this does not pose a problem since the accessory pigments (carotenoids) absorb strongly in the blue region of the spectrum but make no contribution to absorption in the red. Thus determinations of chlorophyll are generally made in the red region of the spectrum.

4.1.1 *Estimating Chlorophyll in Isolated Chloroplasts*

This is the most straightforward case, because it is simple to extract totally the chlorophyll with 80% (v/v) aqueous acetone. It is particularly important to protect extracted pigment from light in order to prevent photo-oxidative loss of the pigments (see below). This is best achieved by covering tubes and other glassware with aluminium foil. A typical procedure is given below.

(i) Take approximately 10 ml of 80% aqueous acetone and add 0.1 ml of chloroplast suspension to it.
(ii) Shake well.
(iii) Sediment the chloroplasts by centrifugation (2000 *g*, 5 min) using centrifuge tubes that are not attacked by the solvent. Take the clear, green supernatant and keep it in a tube covered with foil.
(iv) Re-extract the pellet with a further volume of 80% acetone, centrifuge again and add the supernatant to the supernatant of the first centrifugation. Make up to a known volume (say, 20 ml). (Alternatively, make a single 80% acetone extract and filter it through a Whatman No 1 filter paper to remove extracted chloroplasts and leave an optically clear solution. There should be no trace of pigment on the filter paper).
(v) Using a 10 mm cell in a single beam spectrophotometer, measure the absorbance at 645 and 663 nm or, using a scanning spectrophotometer, obtain a spectrum between 600 and 700 nm.
(vi) Calculate the concentration of chlorophyll (Chl) in the cuvette using the simultaneous equations shown in *Table 1a* (4,5). The extinction coefficients give Chl concentrations in mg l^{-1} ($\mu g\ ml^{-1}$), the conventional way of expressing

Table 1. Equations for Estimating Chlorophyll in Samples Extracted with Organic Solvents.

(a) *Extraction with 80% (v/v) acetone*

Measure the absorbance of the solution at 645 and 663 nm, then:

$$\text{Chl } a = 12.7\,A_{663} - 2.69\,A_{645}$$
$$\text{Chl } b = 22.9\,A_{645} - 4.68\,A_{663}$$
$$\text{Total Chl} = 20.2\,A_{645} + 8.02\,A_{663}$$

(b) *Extraction with methanol*

Measure the absorbance of the solution at 650 and 665 nm, then:

$$\text{Chl } a = 16.5\,A_{665} - 8.3\,A_{650}$$
$$\text{Chl } b = 33.8\,A_{650} - 12.5\,A_{665}$$
$$\text{Total Chl} = 25.8\,A_{650} + 4.0\,A_{665}$$

Chl concentrations in photosynthesis work. The concentrations can be converted to molar knowing that the molecular weight of Chl *a* is 893.5 and that of Chl *b* is 907.5.

As an alternative to using two absorption measurements, it is possible to estimate Chl by using the absorption at 652 nm of an 80% acetone extract, where the extinction coefficients of Chl *a* and *b* are equal, so that:

$$\text{Total Chl} = 27.8\ A_{652} \qquad \text{Equation 15}$$

where the units of concentration are mg l^{-1}. The estimation using 652 nm must be used with care, since it is on a rising shoulder of the 663 nm peak, and inaccuracy in selecting the wavelength can result in significant errors.

The chlorophyll estimates give values for the chlorophyll concentrations in the spectrophotometer cell, and must be corrected for dilution from the original chloroplast sample. In the example cited above, a dilution of 0.1 ml into 20 ml, which is 200-fold, was made. The values given by the equations must therefore be multiplied by 200 to give the correct value. If dilutions are always the same, then the dilution can be incorporated into the extinction coefficient. Hence, for the measurement at 652 nm using a 200-fold dilution,

$$\text{Total Chl} = 5.56\ A_{652} \qquad \text{Equation 16}$$

where the chlorophyll concentration is now in mg ml^{-1} (g l^{-1}).

Table 1 also gives extinction coefficients for methanolic extracts of chloroplasts.

4.1.2 *Extracting Chlorophyll from Intact Tissue*

Total extraction of pigments from intact tissue can be tricky. Acetone or methanol are frequently used for leaves (*Table 1*), and the tissue should be cut into small pieces to allow effective penetration of the solvent, which will be assisted by grinding or macerating the tissue. Small amounts of tissue can be extracted by grinding with solvent in a glass homogeniser, whilst for tough tissues, grinding with the solvent and sand may be necessary.

After grinding or maceration, the tissue and solvent can be separated by centrifugation or filtration, the former technique allowing for re-extraction of the tissue if necessary. To avoid light-induced breakdown of pigments, extracts should be kept in the dark and spectrophotometric analysis should be carried out as soon as possible.

The extraction of chlorophyll from intact algae can also be rather difficult. For green algae, cold or warm (40°C) methanol is successful, although the use of hot solvents can lead to the formation of breakdown products and consequent inaccuracy in estimations. Extinction coefficients for methanol extracts are given in *Table 1b*. The procedure normally involves centrifugation of the algae before solvent extraction: this needs to be taken into account in the dilution factor.

4.1.3 *Breakdown Products*

A major problem with chlorophyll estimations is the formation of breakdown products during and after pigment extraction. Chlorophyll in organic solvents is photolabile, and is quickly and irreversibly bleached. Therefore it is important to exclude light from extracted pigment (*Figure 5*).

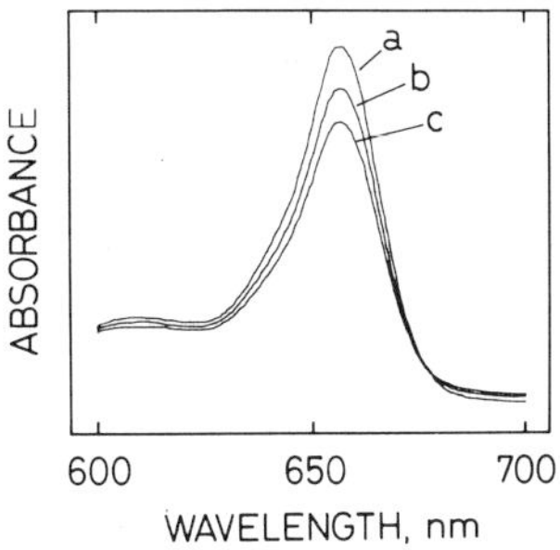

Figure 5. The effect of exposure to light on the absorption of chlorophyll in 80% acetone. Chlorophyll was extracted from a sample of pea chloroplasts (Section 4.1.1) and the absorption spectrum was recorded immediately (**spectrum a**). The sample was then allowed to stand in room light for 10 min (**b**) and 20 min (**c**). Significant loss of absorption is seen.

Table 2. Equations for Estimating Phaeophytin in Samples Extracted with 80% (v/v) Acetone.

Measure the absorbance at 655 and 666 nm, then:

$$\text{Phaeo } a = 20.15\, A_{666} - 5.87\, A_{655}$$
$$\text{Phaeo } b = 31.90\, A_{655} - 13.40\, A_{666}$$
$$\text{Total Phaeo} = 6.75\, A_{666} + 26.03\, A_{655}$$

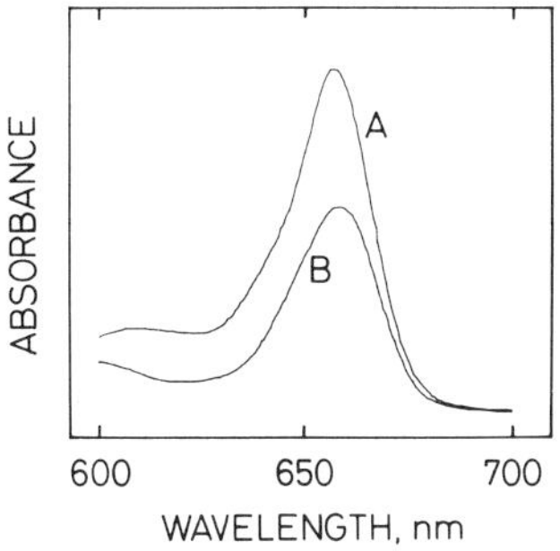

Figure 6. Change in the absorption spectrum of chlorophyll in 80% acetone induced by acidification. (**A**) untreated sample; (**B**) sample acidified with 50 μl 1N HCl. Sample volume 3.5 ml.

Under acid conditions, chlorophyll forms phaeophytin, where the magnesium ion in the porphyrin 'head' group of chlorophyll is replaced by two protons. Phaeophytin can be formed if rather acid tissue is homogenised, although this can be prevented to a certain extent by adding basic substances like $MgCO_3$ or Na_2CO_3 during grinding. Phaeophytin can be estimated spectrophotometrically: an 80% acetone extract can be acidified with saturated oxalic acid in 80% acetone, and the absorbance of the resulting solution is measured at 655 and 666 nm. The extinction coefficients for phaeophytin (Phaeo) in the cuvette are shown in *Table 2*, where the concentrations are in mg l^{-1}. *Figure 6* shows the absorption spectra of an 80% acetone extract of chloroplasts before and after acidification. The degree of phaeophytinisation of an extract may therefore be calculated from the difference between the chlorophyll estimate and the phaeophytin estimate after acidification, since all the chlorophyll will then be phaeophytinised.

4.2 In Vivo Absorption Spectrophotometry

It is an attractive idea to be able to make absorption measurements on intact tissue so that, for instance, accurate chlorophyll assays might be made without first extracting the pigment into organic solvent. The difficulties of working with turbid solutions have been alluded to above (Section 3.2), and in general the difficulties of working with translucent, light-scattering material are not to be underestimated. The same problems which beset working with turbid solutions are found here, that is the increased path-length of the measuring beam and, more seriously, the scattering of the measuring beam out of the direct light path, giving apparent absorption. One component of the scattering is dependent on wavelength, which can pose special problems when a scanning spectrophotometer is used to measure an absorption spectrum.

Another problem arises from the non-uniformity of absorbance in samples: a dilute suspension of algal cells has points of high absorbance (the cells) but the spaces between the cells have negligible absorption. Thus, some light will pass through the sample unattenuated . This is called the 'sieve effect' and is clearly not accounted for by the Beer-Lambert law.

4.2.1 *Dilute Cell Suspensions: the Opal Glass Method*

Taking these problems into account, a method of *in vivo* absorption spectroscopy was proposed by Shibata (6) who recognised that the apparent absorption spectrum of a dilute suspension of algal cells is composed of the two parts noted above: parallel transmitted light which has not been attenuated or scattered by cells, and diffuse transmitted light which has interacted with cells. Reflected light is neglected. In conventional spectrophotometers, the distance between the sample cell and the photomultiplier is generally quite large. The consequence is that the photomultiplier tends to detect predominantly parallel transmitted light, with only a relatively small proportion of diffuse transmitted light. Moreover, the relative proportion of the two transmitted lights depends on the geometry of the spectrophotometer, and will therefore vary from one model to the next. Conventionally measured absorption spectra give very high values for absorption and distort the spectra.

The opal glass method involves using a conventional scanning spectrophotometer, but with the insertion of a diffusing plate next to the sample and reference cells, on the side nearest the photomultiplier. Both parallel and diffuse transmitted light strike the opal glass plate and are completely diffused by it: the same proportion of parallel and diffuse transmitted light that strike the plate is therefore detected by the photomultiplier.

The diffusing plate need not be of opal or ground glass; a perfectly satisfactory plate can be made from a piece of filter paper whose opacity is reduced by being soaked in non-light absorbing oil (typically paraffin oil), drained, then sandwiched between two microscope slides (or quartz plates for work in the u.v.). As Shibata (6) pointed out, the ideal spectrophotometer cell for this type of work is thin and of large cross-sectional area: using conventional 10 mm^2 cells, too large a proportion of diffuse transmitted light may be lost by scattering from the sides of the cell.

4.2.2 *Highly Scattering Samples*

The opal glass method is an empirical procedure for minimising the sieve effect; it is of little value where the sample has already scattered all of the measuring beam. Butler (7) has analysed the absorption properties of highly scattering samples and found that the peaks of the spectrum tend to be flattened. This occurs because the path-length taken by the measuring beam through the sample is wavelength dependent, and is a function of the sample reflectivity. The reflectivity is low where absorbance is high, so the increased path-length effect is least marked where the absorption peaks are seen.

It is difficult to correct for flattening (7). One approach is to bring the sample and photomultiplier as close together as possible, so that the photomultiplier collects light from as large a solid angle as possible. Another approach is to use an integrating sphere. Typically this consists of a hollow sphere, made reflective on the inside by coating it with highly reflective white paint, with the sample placed at the centre. The measuring beam is admitted through a port, and the photomultiplier views the sample through a second port, where it collects the light that the sample has transmitted and reflected. The signal is therefore composed of the light not absorbed by the sample. The absorption spectrum produced by an integrating sphere is thus a mixture of an absorption spectrum and reflectance spectrum (7). Sometimes a port is provided to measure the reflectance spectrum alone.

From this brief consideration, it can be seen that *in vivo* absorption spectroscopy, except for dilute cell suspensions, yields data that are not amenable to simplistic analyses.

4.3 **Estimation of Electron Transfer in Isolated Thylakoid Membranes**

In intact cells the terminal electron acceptor for photosynthetic electron transfer is generally carbon dioxide. When thylakoid membranes are isolated, however, the chloroplast envelope is frequently ruptured, and the soluble components of the chloroplast stroma are washed away into the isolation medium. The electron transfer chain is therefore left without an electron acceptor. As a consequence electron transfer from water is inhibited until an exogenous electron acceptor ('Hill oxidant') is added. Hill oxidants are compounds with suitable redox potentials, enabling them to be reduced at certain sites along the electron transfer chain (Chapter 5). Some Hill oxidants are coloured, and undergo a change in their absorption spectrum in going from the oxidised to the reduced state. For example, oxidised dichlorophenol indophenol (DCIP) is blue, but reduced DCIP is colourless. The rate of electron transfer can therefore be estimated from the rate at which the absorption due to DCIP disappears. Spectrophotometric assay therefore provides an alternative to the oxygen electrode (Chapter 5) for estimating electron transfer rates.

4.3.1 *Electron Transfer Rates using Dichlorophenol Indophenol as Electron Acceptor*

DCIP accepts electrons after photosystem II, and thus measures the flux of electrons through a section of the electron transfer chain from water to the acceptor. A typical assay procedure is given below:

(i) Isolate chloroplast thylakoid membranes by the procedure described in Chapter 2. Store them concentrated in a medium containing 0.1 M sorbitol, 5 mM NaCl,

5 mM $MgCl_2$ and 10 mM 4-(2-hydroxyethyl)-1-piperazine ethanesulphonic acid (Hepes), pH 7.0 (Chapter 5). The precise composition of this medium is not critical, but it should not contain an added reductant, such as ascorbate.

(ii) Add an aliquot of chloroplasts to 3 ml of the same medium to give a chlorophyll concentration of about 25 μg ml^{-1}, and add DCIP to give a final concentration of 50 – 100 μM.

(iii) Place the reaction mixture in a spectrophotometer cell and measure the absorbance at 600 nm in a spectrophotometer, using as a reference a similar reaction mixture but without the DCIP.

(iv) Remove the sample cuvette from the spectrophotometer and illuminate it with actinic light. White light from a slide projector or microscope lamp is suitable, and should give an intensity saturating for the reaction, typically greater than 50 W m^{-2} (200 μmol photons m^{-2} sec^{-1}).

(v) After a short period (say 1 min) shake the sample cuvette and return it to the spectrophotometer, where the absorbance at 600 nm is again read.

(vi) Steps (iv) and (v) are repeated until a linear time course is obtained.

The rate of loss of absorption can be converted to rate of reduction of DCIP by using the extinction coefficient for DCIP, which is approximately 2.06×10^4 M^{-1} cm^{-1} at 600 nm and pH 7.0. Clearly, the reaction will stop when all the added DCIP has been reduced.

There is a disadvantage in using DCIP as an electron acceptor for this type of assay, because it tends to act as an uncoupler of photophosphorylation. Therefore the ratio of uncoupled:coupled electron transfer rates in chloroplasts will be smaller when assayed with DCIP than using other methods.

4.3.2 *Electron Transfer Rates Determined using Other Electron Acceptors*

DCIP is only one of a number of electron acceptors that undergo absorption changes on reduction. Potassium ferricyanide, NADP and ferredoxin have all been used for spectrophotometric assays of electron transfer (Chapter 5). It is easier to use compounds whose absorption change takes place at a wavelength where the absorption by chlorophyll is minimal: from this point of view none of these compounds is as convenient as DCIP. Ferricyanide is relatively frequently used but suffers from the drawback that its extinction coefficient is about one twentieth that of DCIP. Therefore, despite its drawbacks, DCIP is most often used where a simple assay of uncoupled rates of electron transfer through photosystem II is required.

4.3.3 *Continuous Measurement of Electron Transfer*

The experimental technique described in Section 4.3.1 yields discontinuous data. In principle it should be possible to derive continuous data using a modified scanning spectrophotometer, where the sample cell can be illuminated, but the reference cell is kept darkened. Some manufacturers offer modifications to commercial instruments, whilst in other cases home-made modifications have been made. Instruments of this type are discussed in Section 5.2.

4.4 Identification of Pigment Complexes

It is clear that *in vivo* all photosynthetic pigments are present in complexes with protein, rather than as free pigment (Chapter 3). The pigment-protein complexes of photosynthetic bacteria and higher plants are intrinsic membrane proteins which means that they are very hydrophobic, and can only be separated by use of techniques involving solubilisation of the membrane with surfactants (Chapter 3).

The absorption spectrum of a pigment is sensitive to the physical microenvironment that the pigment experiences. In a trivial case, the absorption maximum of a pigment may be expected to vary by a few nanometres if it is suspended in different organic solvents. But in the case of photosynthetic pigment-protein complexes the shift in wavelength can be much more marked. Chlorophyll has an absorption peak at about 663 nm in 80% acetone, but *in vivo* this peak can be red-shifted by up to 40 nm due to the physical interaction between the pigment and the apoprotein.

The degree of red-shifting of the chlorophyll absorption spectrum affords a method of identifying different pigment-protein complexes. The complexes from higher plants and bacteria are most often designated by their absorption maxima in the red region of the spectrum. The analysis can be complementary to biochemical analysis of photosynthetic components.

4.4.1 *Spectrophotometric Analysis of Pigment-protein Complexes in Photosynthetic Bacteria*

Photosynthetic bacteria are particularly amenable to spectrophotometric analysis, since *in vivo* spectrophotometry presents few problems. Subcellular membrane vesicles ('chromatophores') which contain all the apparatus necessary for the light reactions can be isolated from intact cells by mechanical disruption. Chromatophores are very small, and give a suspension which approximates closely to an optically clear solution. Chromatophores can then be treated with detergents to solubilise the membranes and release the pigment-protein complexes which are then separated by conventional methods.

Figure 7 illustrates the absorption properties of chromatophores from a photosynthetic bacterium, and of the constituent pigment-protein complexes. The complexes have been named by their absorption maxima, so that the complex with two peaks at 800 and 850 nm is called B-800-850. It can also be seen that the chromatophore absorption spectrum can be 'reconstituted' by taking various proportions of the pigment-protein complex spectra. As might be expected, the majority of the spectrum is accounted for by the light-harvesting complexes, with only a relatively minor contribution from the reaction centre complex.

4.4.2 *Spectrophotometric Analysis of Pigment-protein Complexes in Higher Plants*

Analogous to photosynthetic bacteria, spectrophotometric analyses of pigment-protein complexes from higher plants is also possible. After fractionation of the pigment-proteins by polyacrylamide gel electrophoresis (see Chapter 3) slices of the gel containing the individual pigment-proteins can be removed and placed in a cell for measurement of

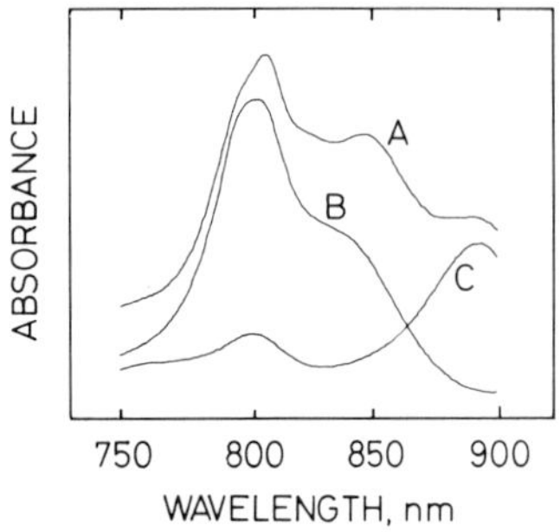

Figure 7. The absorption spectra of chromatophores and pigment-protein complexes isolated from *Chromatium vinosum* strain D. Chromatophores yield the spectrum labelled **A**; spectrum **B** is a mixture of two light-harvesting pigment-protein complexes designated B800-850 and B800-820 from their near i.r. absorption maxima, whilst spectrum **C** is comprised of the B890 light harvesting pigment-protein complex and some reaction centres. Qualitatively it is possible to see how the isolated complexes contribute to the chromatophore absorption spectrum. These data cannot be treated quanitatively as no attempt has been made to normalise the curves. Data courtesy of R.J.Cogdell and D.Dawkins, Botany Department, University of Glasgow.

their absorption properties. Some commercial spectrophotometers have densitometric attachments designed to scan gels, and may have a facility for recording the absorption spectra of bands on gels.

4.4.3 *Specialised Techniques to Improve the Resolution of Spectra*

Two rather specialised techniques may be used to improve the resolution of the minor shoulders in an absorption spectrum. It should be noted that neither of these techniques is normally found in non-specialist laboratories, but they are described for completeness.

(i) *Low temperature spectrophotometry.* Measurement of a spectrum at low temperature (typically liquid nitrogen temperature, 77 K) sharpens the peaks and reveals the details of small shoulders. This is due to a reduction of the thermal excitation of vibrational sub-states. Measurement of low-temperature absorption spectra requires a special attachment to the spectrophotometer provided by some manufacturers that not only maintains the sample at the correct temperature, but also prevents the formation of condensation on cooled windows. Moreover, the frozen sample must remain as nearly optically clear as possible, which can be achieved by using only a very thin layer of sample, or by adding a substance such as glycerol, depending on the type of low-temperature accessory used.

(ii) *Derivative spectra.* Differentiation of an absorption spectrum with respect to wavelength (yielding a spectrum of $dA/d\lambda$) can reveal detail that is not seen in a straightforward absorption spectrum. In fact, second derivative ($d^2A/d\lambda^2$) and fourth derivative ($d^4A/d\lambda^4$) are normally used. The techniques for achieving such spectra are only really practicable when the spectrophotometer is connected to a computer, and digitised spectra in the computer memory can easily be manipulated. Interpretation of derivative spectra of biological materials containing many pigment species is complex and must be undertaken with care (8).

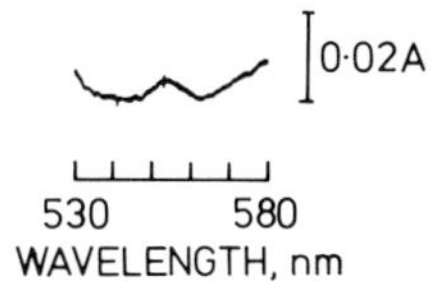

Figure 8. Reduced-minus-oxidised difference spectrum for cytochrome-*f* in chloroplasts. The procedure given in Section 4.5.1 was followed. The spectrum was recorded using a Pye-Unicam SP8000 spectrophotometer: this instrument has the facility for placing the sample and reference cells very close to the photomultiplier, enabling light to be gathered over a large solid angle. The spectrum was recorded about 20 min after addition of oxidant and reductant.

4.5 Difference Spectroscopy

A scanning spectrophotometer can also be used to compare samples, by placing the control sample in the reference beam and the treated sample in the sample beam. This technique can be used to determine the absorbance changes when a sample is subjected to some treatment. For example, cytochromes are complexes of haem and protein which participate in photosynthetic electron transport. With changes in their oxidation-reduction state they show absorption changes in the blue (the 'Soret' bands) and in the green-yellow regions of the spectrum. Cytochrome *f*, for example, shows a bleaching at 554 nm on oxidation. This absorption change can be used to estimate cytochrome *f* in photosynthetic samples (9). Using a scanning spectrophotometer two samples are taken, one is oxidised and placed in the reference beam whilst the other is reduced and placed in the sample beam. The spectrophotometer is then used to record the difference spectrum using, where possible, cell positions close to the photomultiplier tube. The same technique can be applied to other components, like P700 (Chapter 2).

4.5.1 *Reduced-minus-oxidised Difference Spectrum of Cytochrome f*

This absorption change is measured in the presence of the detergent Triton X-100 which removes the interference from cytochrome *b*-559 (9).

(i) Isolate chloroplast thylakoid membranes by the method given in Chapter 2 and resuspend the chloroplasts as a concentrated stock solution in a medium containing 0.33 M mannitol, 1 mM $MgCl_2$, 1 mM $MnCl_2$, 2 mM EDTA, 50 mM phosphate buffer, pH 6.5 and 1% (w/v) Triton X-100.

(ii) Dilute the chloroplasts in the same medium to give a chlorophyll concentration of the order of 170 μg ml^{-1}.

(iii) Using matched cells, place equal volumes of the chloroplast suspension in the sample and reference positions and record a baseline.

(iv) Using 0.5 M stock solutions, add 10 μl of hydroquinone into the sample cell and 10 μl of $K_3Fe(CN)_6$ into the reference cell, giving about 1.5 mM of the added oxidant or reductant.

(v) Record the spectrum from 520 to 580 nm. Correct for the baseline spectrum if necessary, and draw a baseline between 543.5 and 560 nm.

(vi) Estimate the absorption at 554 nm (*Figure 8*). The extinction coefficient for cytochrome *f* is approximately 20 mM^{-1} cm^{-1} at 554 nm.

4.5.2 *Oxidised-minus-reduced Difference Spectrum of P700*

A P700 difference spectrum is difficult to measure in unfractionated chloroplasts (10), but rather simpler in preparations enriched in P700 such as those described in Chapter 2. The principle of the technique is exactly as described in the previous section, except that a chlorophyll concentration of about 25 μg ml^{-1} (giving an absorbance of around 2 at 680 nm) is used and the difference in absorption is measured at 430 nm.

(i) Place two similar samples into matched cells and record a baseline spectrum in the range 410 – 450 nm.

(ii) Oxidise the sample with a small volume of $K_3Fe(CN)_6$, giving 1 mM finally. The extinction coefficient of P700 at 430 nm is about 44 mM^{-1} cm^{-1}.

5. KINETIC SPECTROSCOPY

The spectroscopic techniques described so far have measured absorption in the dark, and in a time-independent manner. But in photosynthetic systems there are a number of transient absorption changes that can be excited by light. It would therefore be very useful to have spectrophotometers in which the absorption of the sample could be monitored at the same time as it was excited either with a brief flash or steady-state light. Such spectrophotometers are quite highly specialised and have generally been available only if constructed in the laboratory. Because of the specialised nature of the equipment, this section does not attempt to give detailed instructions for the use of kinetic spectrophotometers. It does, however, give some design criteria for the simplest single-beam flash spectrophotometer. The remainder of the section is intended to give an understanding of the experimental methods so that results in the literature can be judged critically.

5.1 **The Photosynthetic Events Giving Time-dependent Absorption Changes**

There are four main classes of processes that lead to time-dependent absorption changes. Taking them in order of their characteristic time scales, they are as follows.

(i) *Transient excited states of pigment molecules.* Both chlorophyll and carotenoids can be excited to their triplet states (Section 2.1.2) giving absorption bands which last as long as the triplet lifetime, typically a few microseconds.

(ii) *Oxidation-reduction changes in electron transfer chain components.* This covers a wide range of phenomena from the sub-microsecond photo-oxidation of the reaction centre chlorophylls of photosystems I and II to the slow (millisecond) redox changes of secondary electron transfer components like cytochromes. Almost all electron transfer components show absorption changes on oxidation and reduction, but in some cases (e.g. the iron-sulphur centres) the extinction coefficients are so small, or in relatively unfavourable wavelength regions, that absorption spectrophotometry is not the technique of choice for their study.

(iii) *Electric field across the thylakoid membrane.* A transmembrane electric field across the thylakoid membrane is established as part of the 'high-energy state' that couples electron transport to photophosphorylation (Chapter 6). The transmembrane electric field alters the absorption spectra of some photosynthetic pigments embedded in the membranes, and the resulting light-induced absorption changes act as an *in vivo* voltmeter. The onset of the changes is very rapid but their decay is normally on the millisecond time scale.

(iv) *Scattering*. Chloroplasts tend to scatter light and the degree of scattering depends on the size of the particles. The steps leading to phosphorylation involve swelling and shrinkage of the chloroplasts, which results in scattering changes. The time scale is relatively slow, milliseconds to seconds.

Kinetic spectrophotometers fall into two broad classes, depending on the type of light source used to excite the photosynthetic absorption change which is being monitored. The source can be either (i) a steady-state lamp or (ii) a flash-lamp.

5.2 Spectrophotometers using Steady-state Exciting Light

When using steady-state light to excite photosynthesis, the time resolution of the instrument is limited to a few milliseconds by mechanical considerations (see below), but the time scale of the measurements may range from a few seconds to many minutes. When studying absorption changes that last for more than a few hundreds of milliseconds, it is necessary to compensate for artefactual absorption changes that are due, for example, to light scattering or slight sedimentation of the sample. Compensation is normally made by one of the two following methods.

5.2.1 *Split-beam Spectrophotometers*

Figure 9 is a schematic diagram of a split-beam spectrophotometer, which can be thought of as a modification of a conventional scanning spectrophotometer. The principal differences are as follows.

(i) The sample cuvette can be illuminated with a powerful actinic beam, which excites photosynthesis, whilst the reference cuvette (containing a similar biological sample) remains in the dark. The weak measuring beam is switched so that it passes first through one cell and then through the other, and the amplifier measures

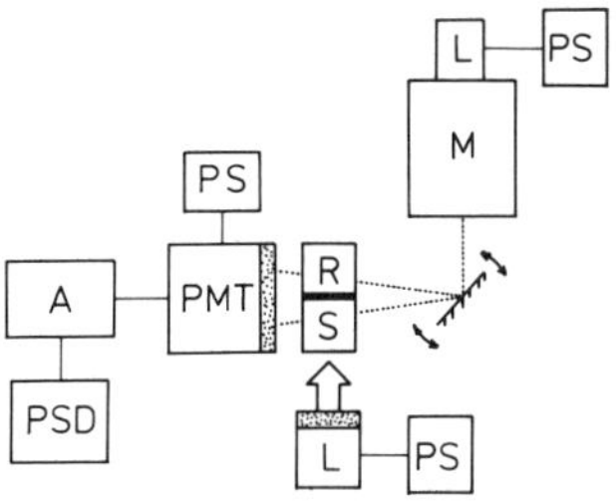

Figure 9. Schematic diagram of a split-beam spectrophotometer. PS, power supply; L, lamp; M, monochromator; R, reference cell; S, sample cell; PMT, photomultiplier tube; A, amplifier; PSD, phase-sensitive detector (lock-in amplifier). The light beam (dotted line) is switched between the reference and sample cells by an oscillating mirror. The sample cell can be illuminated with high-intensity actinic light, whilst the reference cell remains darkened. The actinic light is passed through optical filters (stippled); the photomultiplier tube is protected by a filter which transmits the weak measuring beam but blocks the actinic light. The amplifier is normally a phase-sensitive detector linked to the frequency of the oscillating mirror: this arrangement helps eliminate artefacts arising when a fraction of the actinic beam is detected by the photomultiplier.

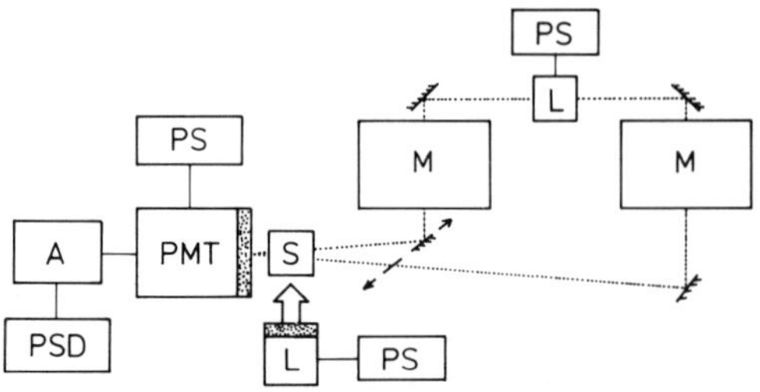

Figure 10. Schematic diagram of a dual-beam double-wavelength spectrophotometer. PS, power supply; L, lamp; M, monochromator; S, sample; PMT, photomultiplier tube; A, amplifier; PSD, phase-sensitive detector. There are two light beams (dotted lines) from the lamp, each of which passes through a monochromator, defining a reference wavelength and a measuring wavelength (see text). An oscillating tuning fork switches first the reference beam and then the measuring beam through the sample cell. The sample is excited with actinic light; the phase-sensitive detector is linked to the frequency of the tuning fork to eliminate artefacts that might arise if the optical filters (stippled) over the actinic lamp and photomultiplier tube are not completely efficient at protecting the photomultiplier from the actinic light.

the difference. The filters over the actinic beam and the detecting photomultiplier are carefully chosen so that the photomultiplier does not detect the actinic light.

(ii) The geometrical layout of the spectrophotometer is designed to allow the cuvettes to be placed very close to the photomultiplier tube so that the light scattered by the sample can be gathered over a large solid angle.

5.2.2 *Double-beam Dual-wavelength Spectrophotometer*

In this type of spectrophotometer there is only one sample cell, but two weak light beams (measuring and reference) that have different wavelengths (*Figure 10*). The method for compensation of artefacts relies on the assumption that the two beams may be chosen to have closely spaced wavelengths where one wavelength measures the absorption change of interest together with the artefactual changes, while the other (the reference wavelength) measures only the artefactual changes, because the absorption change of interest has no magnitude at the reference wavelength (i.e. the reference wavelength is chosen to be at the isosbestic point of the absorption change under study). The signal is therefore the difference in absorption between the two beams.

In both the split-beam and double-beam spectrophotometers, the switching of the light beams is achieved mechanically by an oscillating mirror or tuning fork. The frequency at which the mirror switches the beams gives an estimate of the lower limit of time response of the instrument. Typical values would be of the order of a few milliseconds. Both types of instrument are therefore used for the longer-term light-induced absorbance changes. For shorter times, instruments that employ flash excitation are used.

A typical application of the spectrophotometers employing steady-state light is the estimation of P700 (10). The steady-state instruments have an advantage over flash spectrophotometers in this case, since $P700^+$ is re-reduced very rapidly, necessitating a flash spectrophotometer with a very fast response time. As compared to chloroplasts, better data can be obtained from preparations enriched in P700 (Chapter 2). Photosynthesis is excited with broad-band blue light. P700 measurements are subject to artefacts, particularly that the measuring and exciting beams excite chlorophyll fluorescence at 700 nm, which looks to the photomultiplier like a decrease in absorption. The artefact

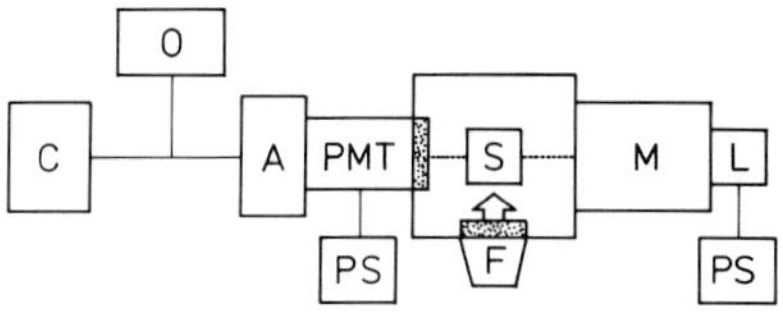

Figure 11. Schematic diagram of a single-beam flash spectrophotometer. The optical design is very similar to a conventional single-beam spectrophotometer (*Figure 3*), but with the addition of a flash lamp (F) and optical filter (stippled) together with an optical filter over the photomultiplier tube (PMT) to block the flash. The output of the amplifier (A) can be passed to an oscilloscope (O) or computer (C). Other abbreviations: L, lamp; PS, power supply; S, sample cell.

can be made smaller by increasing the distance between the sample cell and the photomultiplier, and also by using a narrow bandwidth interference filter over the photomultiplier. For dual-wavelength measurements the reference beam is usually set at 720 nm.

5.3 Flash Spectrophotometers

Flash spectrophotometers are used for measuring light-induced absorption changes whose duration is too short to be measured in a spectrophotometer employing steady-state exciting light. The design is relatively simple, and as *Figure 11* shows, it is essentially a modification of a conventional single-beam spectrophotometer (cf. *Figure 3*), and is very similar to the classical flash photolysis equipment developed by Norrish and Porter.

5.3.1 *Design Criteria for a Single-beam Flash Spectrophotometer*

Single-beam spectrophotometers, although specialised pieces of equipment, are fairly easily constructed. The individual components, however, need to meet some design criteria which are outlined below.

(i) *Generating the measuring beam.* The lamp which provides the light for the measuring beam needs (a) to have an extremely stable output, and (b) to have sufficient intensity in the wavelength range of interest. For the visible and near i.r., a quartz-halogen slide projector lamp or car headlamp is satisfactory. In the latter case a cheap d.c. supply can be provided by a car battery and rheostat but it is more convenient to use a fully smoothed d.c. supply run from mains electricity. Attention should be paid to the ripple of the supply, which should be of the order of one part in 10^5 or less. The bulb may need to be cooled with a fan, but even fans designed for computers will slightly shake the apparatus, and hence make the signal noisy.

(ii) *Monochromator.* A monochromator is an expensive piece of equipment, and it could be replaced (at the cost of greatly reduced flexibility of use) by a fully-blocked interference filter (Appendix II). But a monochromator with variable-width entrance and exit slits offers many advantages. The intensity and spectral bandwidth of the measuring beam can be altered easily, as can the wavelength. A relatively simple grating monochromator, blazed at the wavelength most frequently used is adequate.

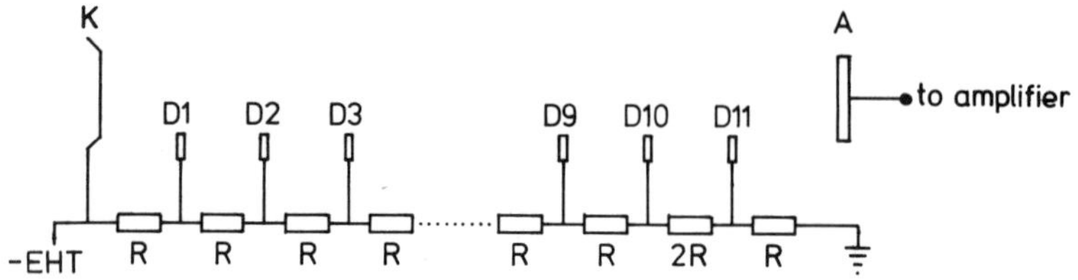

Figure 12. Circuit for the high-voltage supply to a photomultiplier tube. The tube comprises a photocathode (K), several dynodes (D) and an anode (A). The high voltage (−EHT) is supplied to the dynodes via a resistance chain, where the value of R is typically 100 kΩ. Values of the high voltage might be between −500 and −1200 V, depending on the characteristics of the individual photomultiplier tube and the gain required. The output from the anode is a current: this is best treated with a current-to-voltage converter and amplifier (*Figure 13*).

(iii) *Flashlamp.* The duration of the flash used to excite photosynthesis should be short enough so that the photochemical centres are excited only once during the flash. It should also be intense enough to excite all the centres, even when a large fraction of the lamp's spectral output is filtered. Xenon-filled stroboscope lamps with a duration of about 5 μsec have been found satisfactory.

(iv) *Photodetector.* The photodetector needs to have a fast response (at least twice as fast as the most rapidly varying change it is expected to measure), high sensitivity and a spectral response that covers the wavelengths of interest. Despite the advent of photodiodes, photomultiplier tubes are still favoured as the photodetector, since their response times are a few tens of nanoseconds, their gain can be very high, and their spectral response can be chosen as a function of the materials used to make the photocathode. For example, if one photomultiplier tube is to be used both for absorption and chlorophyll fluorescence, an extended S20 photocathode is useful. The real advantage of photomultipliers, compared to photodiodes, is their large photosensitive area (typically 50 mm diameter). Several companies manufacture suitable photomultipliers, including EMI Electron Tubes (Hayes, Middlesex.), Mullard (London) RCA (United States) and Hamamatsu (Japan).

The disadvantages of photomultipliers are 2-fold. Firstly, they are expensive, and need a thoroughly smoothed high voltage supply. Secondly, suitable circuits need to be constructed to provide the correct high voltage to the component parts of the photomultiplier and to amplify the output signal. *Figure 12* shows the photomultiplier supply circuit, whose function is to deliver a gradually varying voltage to each of the dynodes (electron-emissive surfaces) in the dynode chain. The gain of the photomultiplier is highly dependent on the value of the voltage difference between the dynodes. An amplifier for the signal from the anode is shown in *Figure 13*. It consists of a current-to-voltage converter (the photomultiplier produces a current) and amplifier, together with a resistor-capacitor (RC) circuit for eliminating high-frequency noise generated by the photomultiplier. It also incorporates a 'back-off' circuit, which is used to eliminate the d.c. component of the absorption changes so that the relatively small light-induced change can be amplified satisfactorily.

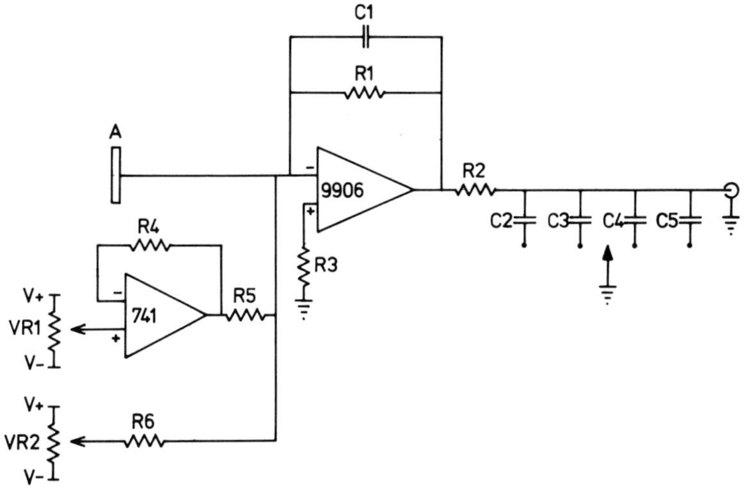

Figure 13. Circuit which takes the current output from the anode (A) of a photomultiplier, passes it through a current-to-voltage converter and then amplifies the resulting voltage and smooths it with an RC filter network. Typical values for the components are: R1 = 100 kΩ, R2 = 1 kΩ, R3 = 1 kΩ, , R4 = 10 kΩ, R5 = 10 kΩ, R6 = 100 kΩ, VR1 = 10 kΩ, VR2 = 10 kΩ, C1 = 10 pF, C2 = 1 nF, C3 = 10 nF, C4 = 100 nF, C5 = 1 μF. The variable resistor VR2 is used to set the zero of the amplifier such that a zero current input results in zero voltage output: once set it need not be further adjusted. VR1 adjusts the variable offset (or back-off) of the amplifier. Switching capacitors C2 to C5 into circuit gradually increases the characteristic time of the RC filter, eliminating more high-frequency noise, although at the same time introducing the possibility of distorting high-frequency components of the signal (see text). For spectrophotometry using a single-beam spectrophotometer such as that shown in *Figure 11*, the variable back-off VR1 would be used to eliminate the steady-state component of the absorption signal so that the small flash-induced component can be amplified satisfactorily. For fluorescence measurements the back-off would not be used.

(v) *Filters*. The filters over the photomultiplier must transmit the measuring beam, but block the flash; the filter over the flash must absorb wavelengths liable to be detected by the photomultiplier, but transmit a range of wavelengths which the sample can absorb. The filters are typically band pass or cut-off, the latter only transmitting light above a critical wavelength (Appendix II).

(vi) *Signal capture.* The simplest way to capture data is to trigger an oscilloscope slightly before the exciting flash is fired. Conventional oscilloscopes can be used with a camera, but storage oscilloscopes are more versatile. It is far more convenient, however, to use a digital storage oscilloscope or transient recorder to capture the signal, since there is then the possibility of manipulating the data.

(vii) *Signal averaging.* For many flash-induced signals, the signal-to-noise ratio is not good enough if one single transient is recorded. An improvement in the ratio can be achieved by digital signal averaging, where a number of similar signals is summed. Because the noise is random in time, it tends to cancel out, but because the signal occurs at the same relative position in the trace all the time, it tends to be reinforced. The signal-to-noise ratio improves as $n^{-1/2}$, where n is the number of signals that is summed. Digital signal averaging has some drawbacks, however. It is necessary to have a computer to time accurately and to trigger all the events in making the measurements; the experimental system must be one that can be stimulated repeatedly and give the same response, and the sample must not degrade during the measurements.

5.3.2 *An Example of using a Single-Beam Flash Spectrophotometer*

A single-beam flash spectrophotometer can be used to measure light-induced absorption changes associated with all the events outlined in Section 5.1. This section will describe the general experimental protocol used for experiments with such an instrument, and take measuring the field-indicating absorption change in chloroplasts as an example.

(i) *Filters*. The field-indicating absorption change in chloroplasts has a maximum at about 520 nm, in the green region of the spectrum, which means that the exciting flash must be in the red region. The flash is therefore passed through red filters (e.g. Kodak Wratten 92), and the photomultiplier is protected with blue-green transmitting filters which absorb red light (e.g. Corning 4-76 and 4-96).

(ii) *Measuring beam.* Whilst the choice of measuring wavelength is very important (see below), the intensity of the weak measuring beam can also affect the measurements. If the intensity is too low, the signals are very noisy. But if the intensity is too high, the measuring beam has some actinic effect, and keeps a fraction of the reaction centres 'closed'. The signal is then smaller than would be expected. It is usual to determine the intensity of the measuring beam empirically: at 520 nm in the green region about 50 nmol photons m^{-2} sec^{-1} (1 μW cm^{-2}) has been found to be suitable, but chloroplast absorption is relatively low in this spectral region.

(iii) *Flash intensity and frequency.* The flash should be saturating; this can be tested by making a flash intensity curve using neutral density filters over the flashlamp. Whether or not the flash is saturating will be related to the chlorophyll concentration of the sample, normally up to 25 μg ml^{-1} for measurements in this spectral region. In experiments employing signal averaging, the flash repetition rate should be chosen so that the system relaxes completely before the next flash is fired. If the flashes are fired too rapidly, the extent of the signal will be decreased. Repetition rates can be quite high for some absorption changes associated with redox reactions, but for those involving ion flux (like the electrochromic absorption change) $0.5-1$ Hz may be more suitable – the exact rate can be determined empirically.

(iv) *Time scale and smoothing the data.* The time base of the recording device will of course be chosen so that the interesting features are captured, but will not be so long that unwanted changes in absorption (e.g., due to scattering or sedimentation) are present. A time scale of 500 msec is a realistic upper limit. If there is a good deal of high-frequency noise, then the RC low-pass filter (*Figure 13*) may be employed to eliminate high-frequency noise, but to allow the signal through, this having a lower frequency. There is the danger of using an RC filter with too low a frequency, which results not only in the elimination of the noise, but also in the distortion of the signal. Again empirical evaluation of a suitable RC filter can be used, with the RC filter of lowest frequency which does not distort the signal being chosen. For a 200 msec sweep, an RC filter with a characteristic time of less than 0.1 msec is normally satisfactory, except in cases where rapid components of the signal are present.

(v) *Characterising the absorption change.* The single-beam flash spectrophotometer has obvious limitations, in particular that the absorbance changes at any wavelength may be due to more than one process, and it is necessary to distinguish the change of interest. One criterion is kinetic — the carotenoid triplets mentioned above decay in a few μsec — and another is spectral. The light-minus-dark difference spectrum of an absorption change can be used to characterise it, and is constructed by taking point-by-point measurements over the wavelength range of interest. In addition some absorption changes are sensitive to certain inhibitors: in the case of the field-indicating absorption change, it is very sensitive to uncouplers and ionophores, so that the uncoupler-sensitive change at 520 nm is taken to be the field-indicating absorption change.

5.3.3 *Improvements to Single-beam Flash Spectrophotometers*

Better resolution of some absorption changes can be made to some extent by using the instrument as a kind of dual-wavelength flash spectrophotometer, and making a signal measurement at one wavelength, followed by a reference measurement at a second wavelength (the isosbestic point for the absorption change under study) and subtracting one signal from the other. Data manipulations of this type are most easily carried out by a laboratory computer attached to the spectrophotometer (11). The double-beam dual-wavelength spectrophotometer uses another method to address the problem raised at the end of Section 5.1, and is essentially two single-beam flash spectrophotometers in a single instrument, enabling two weak measuring beams at different wavelengths to be used simultaneously. The signals from the two photomultipliers can be used independently, or the difference taken. In the latter case, extraneous absorption changes in the signal beam can be eliminated.

6. MEASUREMENT OF FLUORESCENCE

Consideration of the practical aspects of fluorescence measurements in relation to photosynthesis will be confined to fluorescence emission from chlorophylls which occurs at wavelengths between 650 and 800 nm. Measurement of fluorescence emission from photosynthetic systems is relatively simple compared with the measurements of light absorption and this has contributed significantly to the popularity of chlorophyll fluorescence as a probe of the organisation and functions of the photosynthetic apparatus. There are many excellent spectrofluorimeters available commercially, which are often expensive and designed for specific applications. However, an extensive range of fluorescence measurements can be made on simple instruments that can be readily and cheaply constructed in the laboratory. Fluorescence is usually generated by exciting the sample with radiation produced from a white light source (e.g. xenon, quartz-halogen) and transmitted through either a monochromator or glass filters (see Appendix II) to remove emitted wavelengths which are similar to those of the chlorophyll fluorescence to be measured. As with absorption spectroscopic measurements (Section 5.3.1) the light source should provide a stable light output and ideally be driven by a smoothed d.c. power supply run off mains electricity. When using light sources which generate large quantities of heat, the monochromator or glass filters should be protected from the heat by the use of a heat-reflecting filter (see Appendix II). Alternatively, monochromatic excitation radiation can be obtained by using lasers; low power (10 mW and below) helium-neon lasers, which have a major emission at 633 nm, and helium-cadmium lasers, which have a major emission at 441 nm, are commercially available and readily incorporated into 'home-made' spectrofluorimeters. Besides the major emission bands these lasers do emit at other discrete wavelengths and these minor emission bands should be prevented from reaching the photodetector by the use of blocking filters (see Appendix II), which only allow transmission of the major emission band. Since lasers constitute a potentially hazardous radiation source for users, it must be emphasised that they should only be used under well-defined safety conditions, which will be specified by a laboratory's administrative authority. Fluorescence emission from a sample is generally passed through either a monochromator or an interference filter (see Appendix II) to remove reflected and scattered excitation radiation before being passed to a photon detecting device. Photomultipliers or photodiodes can be used to detect fluorescence. Photomultipliers have the advantage of being an order of magnitude more sensitive than photodiodes, however they have the disadvantages of being considerably more expensive and bulky than photodiodes and requiring a high voltage power supply (500 – 1000 V) for operation (see also Section 5.3.1). A suitable photomultiplier for chlorophyll fluorescence spectroscopy is the extended S20 photocathode. Photodiodes are small, cheap and can operate off low voltage power supplies, such as dry batteries. The electrical signal from the photomultiplier or photodiode can be measured directly across a fixed resistance by a potentiometric pen recorder, an oscilloscope or other suitable voltage measuring devices. Alternatively, the signal can be passed to an amplifier, as described in Section 5.3.1, if more precise control on the magnitude of the electrical signal being passed to the recording device is required. When constructing a fluorimeter it is essential to ensure that the interfaces of the component parts are light-tight, if the instrument is to be used in subdued laboratory light and not in a darkroom. The intensity of fluorescence emission from photosynthetic systems is generally very small compared to the magnitude of background laboratory light, thus if apparatus is not light-tight laboratory light will seriously interfere with the measurement of the fluorescence. Interfaces between spectrofluorimeter components can be made light-tight by the use of non-reflective black cloth and/or black insulation tape. Details of the theory and operation of components of a spectrofluorimeter, for example,

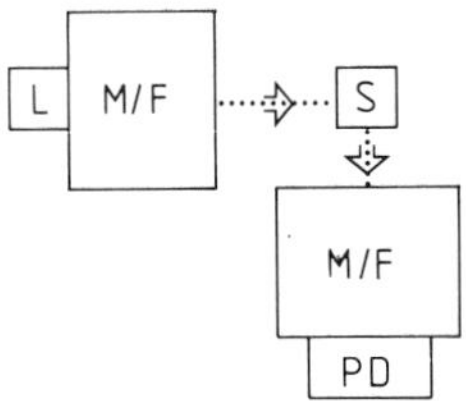

Figure 14. Apparatus for measurement of fluorescence emission from photosynthetic systems in solution. L, light source; M/F, monochromator or optical filter; PD, photodetector; S, sample.

monochromators, interference filters and photomultipliers, are beyond the scope of this chapter and reference (12) should be consulted for such information.

6.1 **Photosynthetic Systems in Solution**

Fluorescence emission from chlorophyll solutions, preparations of isolated photosystem particles, isolated thylakoids, protoplasts and single cell suspensions is generally measured at 90° to the exciting beam (*Figure 14*). The cuvette holder should be designed to allow irradiation of the whole sample and the photodetector should be kept as close to the sample as possible in order to maximise the amount of emitted radiation collected. For measurements from particulate samples, samples should be stirred continuously throughout to avoid a reduction in the fluorescence monitored that would be produced by particle sedimentation. Electromagnetic stirring units, such as those used with oxygen electrodes, are cheap and convenient to use. If stirring facilities are not available particle sedimentation can be prevented by suspending the sample in a solution of a similar specific gravity to that of the sample; 20 – 30% (v/v) glycerol is commonly used for chloroplast preparations and 7% (w/v) Ficoll (mol. wt. 400 000) can be used for single cell algae. When using solutions, such as glycerol, to prevent particle sedimentation the possible osmotic effects of the solution on the fluorescence emission characteristics of the sample should be considered.

6.2 **Leaf Tissue**

Chlorophyll fluorescence from leaf tissue can readily be measured using a conventional spectrofluorimeter and a modified sample cuvette, which holds the leaf surface at 45° to both the exciting beam and fluorescence detector. Generally, the leaf tissue is placed between two pieces of glass, although this is not to be recommended since it modifies markedly the gaseous environment of the leaf tissue, which can affect the fluorescence emission characteristics. A more convenient way to measure fluorescence from leaf tissue is using a bifurcated glass optic fibre to transmit excitation to and fluorescence from the leaf surface (*Figure 15*). Ideally, the glass fibres of the two branches of the optic fibre should be randomly mixed in the common end to ensure an even distribution of excitation radiation to the leaf tissue. If the glass fibres are not randomly mixed, the use of a glass or Perspex rod in conjunction with a piece of ground glass to transmit radiation from the common end of the fibre optic to the leaf will serve to scatter the excitation radiation and decrease the heterogeneity of excitation of the leaf sample (*Figure 15*). In many physiological studies it is essential to enclose the leaf tissue in a chamber in which the atmosphere and leaf temperature can be controlled. Fluorescence emission characteristics of leaf tissue can be modified markedly by changes in atmospheric CO_2 and O_2 and temperature. The gaseous environment of leaf tissue can be controlled in a sealed chamber by the use of a carbonate-bicarbonate buffer (13). Chambers through which gas of known composition can be passed are ideal for studies of leaf fluorescence and are also commercially available: we have found the LD2 leaf disc oxygen electrode chamber (Hansatech, King's Lynn) particularly useful for fluorescence studies.

When measuring fluorescence from leaf samples using optic fibres as described above,

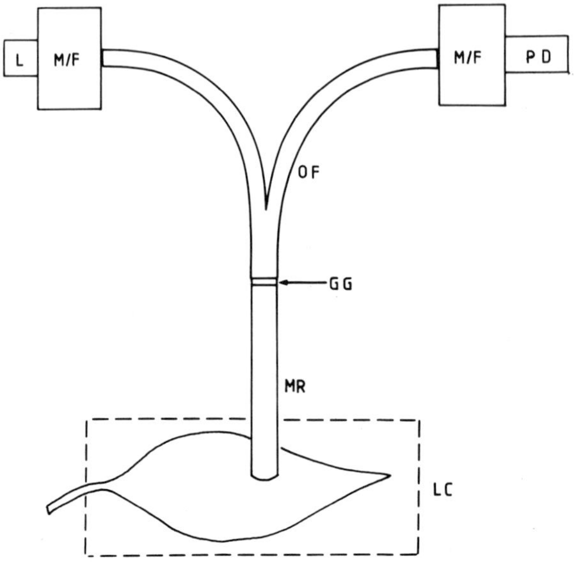

Figure 15. Apparatus for measurement of fluorescence emission from leaf tissue. GG, ground glass disc; L, light source; LC, leaf chamber; M/F, monochromator or optical filter; MR, glass, perspex or quartz light mixing rod; OF, optic fibre; PD, photodetector. The light mixing rod, ground glass disc and optic fibres should be encased in light-tight tubes.

the fluorescence emission detected is not representative of the entire population of leaf cells. Cells at the top of the leaf will contribute considerably more to the measured fluorescence than those in the mid- and lower-leaf regions. Moreover, different wavelengths of excitation radiation will penetrate the leaf tissue to different degrees; for example, blue radiation, which is strongly absorbed by chlorophyll, will penetrate the leaf less efficiently than green radiation, which is weakly absorbed by chlorophyll. The interpretation of fluorescence measurements from leaf tissue can be further complicated by the fact that fluorescence emitted below 700 nm will be strongly absorbed by chlorophyll within the leaf whilst emissions above 700 nm will be considerably less absorbed. Thus fluorescence emissions below 700 nm will be appreciably attenuated relative to those at longer wavelengths as they pass through the leaf. This phenomenon of fluorescence re-absorption, often termed 'self-absorption', is considered in more detail in Section 7.1.

7. APPLICATIONS OF FLUORESCENCE SPECTROSCOPY TO PHOTOSYNTHESIS

7.1 **Emission Spectra**

A fluorescence emission spectrum is usually measured using excitation radiation produced from a white light source in conjunction with a glass filter or a monochromator, and passing the emitted radiation from the excited sample through a scanning monochromator to determine the fluorescence emissions throughout a selected wavelength range (generally 650 – 800 nm). The size of the entrance and exit slits of the scanning monochromator play an important role in determining the spectral resolu-

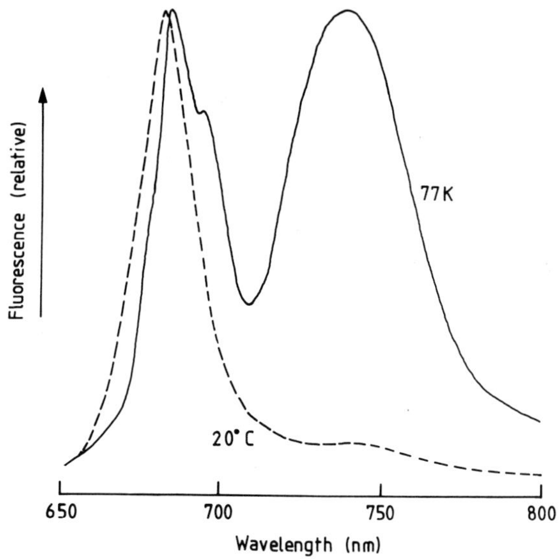

Figure 16. Fluorescence emission spectra of isolated, broken pea chloroplasts at 20°C (– – – –) and 77 K (———). The chloroplasts were suspended in a medium containing 5 mM $MgCl_2$, 10 mM KCl, 15 μM DCMU, 50 mM Hepes at pH 7.6 to a final chlorophyll concentration of 10 μg ml^{-1}. The chloroplasts were excited with a photon flux density of 100 μmol m^{-2} sec^{-1} of 440 nm radiation, having a 10 nm bandwidth, which was produced from a xenon light source and high irradiance monochromator.

tion for the fluorescence transmitted through the monochromator; see reference (12) for theoretical discussion. In practice, for fluorescence spectral studies the emission slit should be set at the minimum aperture to achieve maximum spectral resolution but yet allow sufficient radiation to reach the photodetector to produce a satisfactory signal-to-noise ratio for the detector output. In studies of fluorescence kinetics where quantitative signal analysis is often required, rather than high spectral resolution, wider slit widths can be used. The excitation radiation must not contain the same wavelengths as those emitted by the sample and monitored by the photodetector. At 20°C the emission spectrum from higher plant chloroplasts (*Figure 16*) has a maximum at about 680–685 nm with a small shoulder at about 740 nm. The majority of the emission in this spectrum is from chlorophylls associated with photosystem II; those associated with photosystem I make only a minimal contribution. However, on freezing in liquid nitrogen to 77 K (see Section 4.4.3) three maxima are resolved in the spectrum at about 685, 695 and 740 nm (*Figure 16*). The 685 and 695 nm bands are attributed to antennae chlorophylls of photosystem II; both the light-harvesting chlorophyll-protein complex II (LHCII) and photosystem II core complex contribute to the 685 nm band, whilst the 695 nm band is attributed to pigments of the photosystem II core complex (14). The 740 nm band is generally considered to be emissions from photosystem I chlorophylls, although photosystem II chlorophylls may make a contribution. Photosystem I antennae chlorophylls are also thought to have a major emission in the 720 nm region (14). The accurate allocation of fluorescence emissions at given wavelengths to specific components of the photosynthetic apparatus is currently a contentious issue; the situation is likely to be considerably more complex than suggested above. However, despite

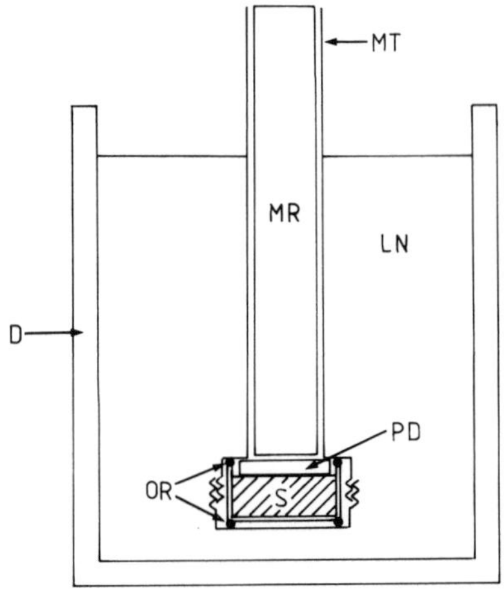

Figure 17. Modification to the optic fibre spectrophotometer, shown in *Figure 15*, to allow measurement of fluorescence from samples frozen to 77 K. The light mixing rod is connected to the ground glass disc and optic fibre (*Figure 15*) using a light-tight connector. D, dewar flask; LN, liquid nitrogen; MR, light mixing rod; MT, metal tube; OR, O ring; PD, perspex or quartz disc; S, sample.

such difficulties the 77 K spectrum has been of considerably more use than the 20°C spectrum for studying the distribution of excitation energy within the photosynthetic apparatus. Many commercially available spectrofluorimeters offer an optional attachment for measurement of samples at temperatures below 0°C. However, an optic fibre spectrofluorimeter (see *Figure 17*) can be readily modified for measurements at 77 K as shown in *Figure 17*. A metal sample cuvette holder should be constructed that attaches by a screw thread to a metal tube which contains the light-mixing rod. The chloroplast sample or leaf tissue can be placed in a polyethylene sample holder, which is then sealed into the metal cuvette holder using a suitable 'O' ring(s). The metal sample cuvette and light mixing rod is then plunged into a Dewar flask containing liquid nitrogen. The Dewar flask should be sufficiently large to ensure that the evaporation of liquid nitrogen does not excessively deplete the volume of liquid nitrogen bathing the cuvette holder. Samples should be immersed in liquid nitrogen for a minimum of 5 min prior to making spectral measurements in order to ensure that a temperature of 77 K has been reached and maintained.

The ratio of emissions at 685 (F685) or 695 nm (F695) relative to that at 740 nm (F740) can be used to examine changes in the relative distribution of excitation energy between the two photosystems, however, care must be taken in its use. The relative magnitude of F685 (or F695) to F740 in a spectrum will be dependent upon the chlorophyll concentration within the sample. Consideration of the absorption spectrum of a chloroplast preparation shows that fluorescence emissions below 700 nm can be re-absorbed by the sample, whilst emissions of the 740 nm band will not be. Thus, as the chlorophyll concentration of the sample is increased, the F740/F685 ratio will increase as shown in *Figure 18*. A chloroplast suspension containing below 10 μg chlorophyll ml^{-1} should be used to minimise 'self-absorption' artefacts. Such chlorophyll concentrations will produce a 685 nm emission of convenient magnitude

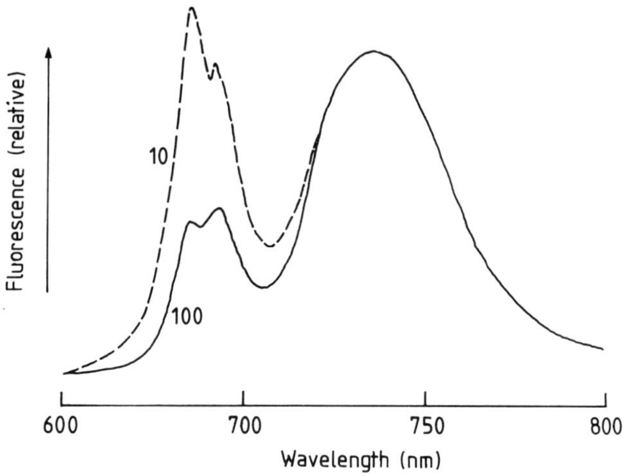

Figure 18. Fluorescence emission spectra of isolated, broken pea chloroplasts containing 10 and 100 μg chlorophyll ml^{-1} at 77 K. The chloroplast resuspension medium and excitation conditions were as described in the legend of *Figure 16*.

for comparison with the 740 nm emission; at chlorophyll concentrations above 10 μg ml^{-1} the 685 nm emission is severely reduced relative to that at 740 nm. Chlorophyll concentrations below 5 μg ml^{-1} may produce 'noisy' signals, which result in poor spectral definition.

On freezing to 77 K samples are often not optically homogeneous due to cracking of the ice. Suspension of chloroplast preparations in glycerol will reduce this problem, however the osmotic effects of glycerol on the photosynthetic apparatus must be considered in the context of your experiment.

In order to determine whether changes in the ratio F740/F685 are due to a change in F740 or in F685 it is essential to obtain reproducible values of F740 and F685. This is achieved by the addition of an internal fluorescent standard to the chloroplasts. For a chloroplast suspension containing 10 μg chlorophyll ml^{-1}, 2 μM fluorescein (sodium salt) provides a good internal standard. Fluorescein emits maximally at 535 nm (*Figure 19*) and by standardising the emission from samples on this peak, it is possible to compare accurately the emission peak values from differently treated chloroplast samples. An example of this technique is given in *Figure 19* where addition of 5 mM $MgCl_2$ to broken chloroplasts enhances the 685 and 695 nm emissions but reduces that at 740 nm.

Fluorescence emission spectra can be measured from leaf samples (see Section 6.2) and spectral ratios determined; however, it is difficult to interpret the significance of differences in the ratios between leaf samples due to possible differences in the optical properties and chlorophyll contents of the leaf samples. It is always preferable to isolate chloroplasts from leaf tissue to make such spectral comparisons, although this is clearly not practicable in experiments where the changes in the fluorescence emission characteristics of the chloroplasts *in vivo* may be modified significantly by the isolation procedures.

Generally, emission spectra are not corrected for changes in the sensitivity of the

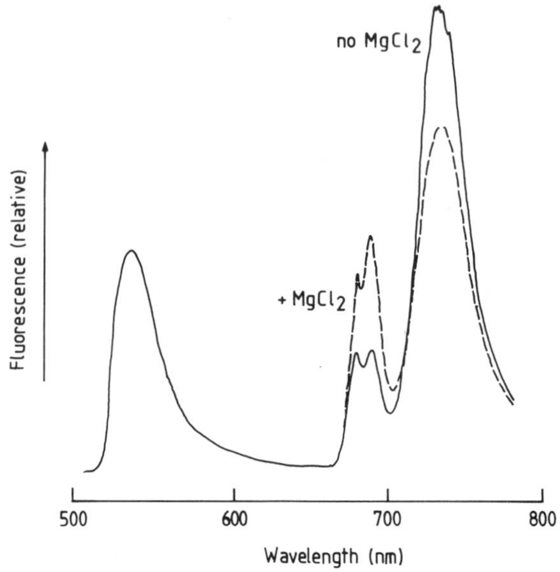

Figure 19. Fluorescence emission spectra of dark-adapted, isolated, broken spinach chloroplasts containing 10 μg chlorophyll ml^{-1} and 2 μM fluorescein (sodium salt) in the presence and absence of 5 mM $MgCl_2$. The spectra were normalized on the emission maxima of fluorescein at 535 nm. The samples were excited with 4 W m^{-2} of 480 nm radiation of 20 nm bandwidth. Full details of the experimental protocol are given in reference (15), from which the figure was redrawn.

fluorescence detector with changes of wavelength since they are only being used for comparative purposes. Photodetectors do not have a constant sensitivity to photons as a function of wavelength, although this may be the case over restricted wavelength ranges. Spectral response curves for photodetector sensitivity can be obtained from the suppliers and manufacturers of photodetectors. If absolute values of fluorescence emission are required (although they are not normally) then the emission spectrum monitored by the photodetector must be corrected using specific wavelength sensitivity values taken from the spectral response curve for the photodetector.

7.2 **Excitation Spectra**

Excitation spectra are measured by exciting samples with radiation transmitted through a scanning monochromator over a given wavelength range and monitoring fluorescence emission at a fixed wavelength. Excitation spectra provide information on the pigments that contribute to the fluorescence emission and are particularly useful in the study of excitation energy transfer from accessory pigments to the chlorophylls associated with photosystems I and II. It is important when examining such energy transfer to ensure that the intensity of the excitation beam is considerably below that required to saturate the pigment bed from which the monitored fluorescence is emitted. Under saturating excitation light, contributions of energy transfer from accessory pigment beds to the monitored fluorescence will be minimal. *Figure 20* shows excitation spectra between 350 and 540 nm for 685 nm emission at 20°C from wild-type barley chloroplasts and from chloroplasts of a barley mutant which lacks chlorophyll *b*. The mutant chloroplasts do not have a major peak at 470 nm compared with those of the wild-type, which is

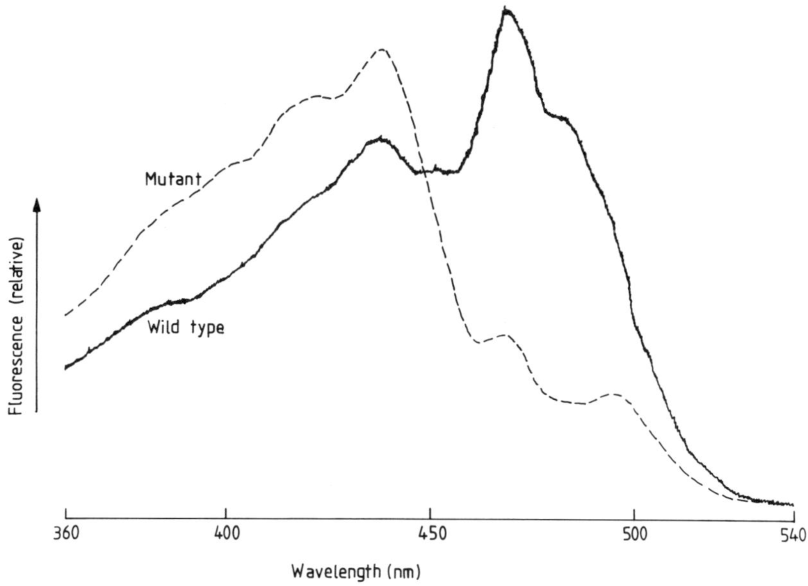

Figure 20. Excitation spectra of 680 nm fluorescence emission of broken chloroplasts of wild-type barley and a mutant lacking chlorophyll *b* (chlorina-f-2) at 20°C. The spectra were measured using a Baird Atomic Fluoripoint spectrofluorimeter on samples containing 10 μg chlorophyll ml^{-1}. The chloroplasts were suspended in the medium described in the legend of *Figure 16*.

the Soret band absorption maximum for chlorophyll *b*. These data demonstrate that light absorbed by chlorophyll *b* makes a major contribution to the photosystem II fluorescence emission in the wild-type but not in the mutant chloroplasts.

If an absolute excitation spectrum is required then the spectrum must be measured using a constant photon flux density to excite the sample at all of the excitation wavelengths used.

7.3 Fluorescence Kinetics

Analyses of the kinetics of fluorescence transients induced from photosynthetic systems by the addition of light or chemical treatment can provide valuable information about the organisation and functioning of the photosynthetic apparatus.

7.3.1 *Measurement of Fluorescence Kinetics*

The kinetics of changes in fluorescence emission from photosynthetic systems in solution or from leaf tissue can be monitored using the apparatus shown in *Figures 14* and *15*, respectively. The electrical signals from the photodetector must be captured on a data recording system, which is capable of resolving the fluorescence transient of interest. Transients occurring over periods greater than a second can be recorded directly onto a good quality potentiometric recorder. More rapid transients should be captured using a device such as a digital oscilloscope, transient recorder or a microcomputer with an analogue-to-digital signal converter. Such devices allow rapidly recorded data to be plotted out slowly on a potentiometric pen recorder or graph plotter after capture. Light-induced fluorescence transients are generated by opening a shutter placed bet-

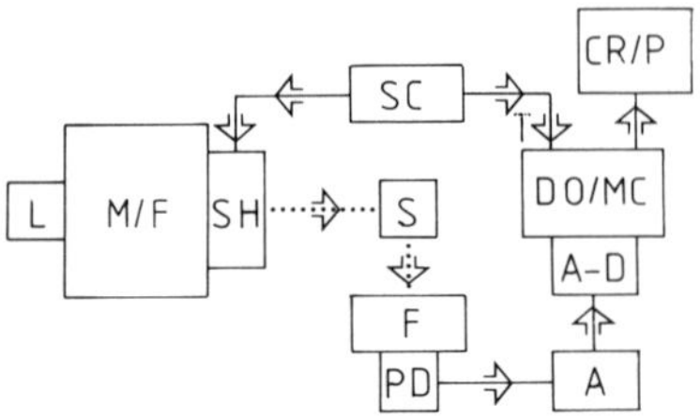

Figure 21. Apparatus for the measurement of fluorescence kinetics from photosynthetic systems in solution. A, amplifier; A-D, analogue-to-digital signal converter; CR/P, chart recorder or plotter; DO/MC, digital oscilloscope or microcomputer; F, optical filter; L, light source; M/F, monochromator or optical filter; PD, photodetector; S, sample; SC, shutter control unit; SH, shutter; T, trigger input.

ween the excitation light source and the sample. Electromagnetically operated shutters are preferable to manually operated, mechanical shutters because of their more rapid opening times and their facility for the electrical triggering of the data storage device at the onset of shutter opening. However, manually operated shutters can be successfully used for transient studies in many experimental applications, provided they are used in conjunction with a data storage device with an internal triggering facility. A typical apparatus for measurement of rapid fluorescence transients from photosynthetic systems in solution is shown in *Figure 21*.

7.3.2 *Fluorescence Induction Curves*

(i) *Primary photochemistry.* At 77 K, or in the presence of 3-(3,4-dichlorophenyl)-1,1-dimethylurea (DCMU) at 20°C, electron transfer from the primary electron acceptor of photosystem II, Q_A, to a secondary quinone acceptor, Q_B, is prevented; thus on exposure of dark-adapted photosynthetic systems to excitation under such conditions, the fluorescence kinetics associated with the reduction of Q_A can be examined. Treatment of photosystem II particles and isolated chloroplasts with 10 – 15 μM DCMU completely inhibits electron transport from Q_A to Q_B, however 50 μM DCMU should be used in experiments with protoplasts and single cells to ensure efficient penetration of DCMU to the sites of action on the thylakoids. Such samples in solution should be dark-adapted for 5 min prior to addition of DCMU and exposure to actinic radiation. Similar experiments can be performed with leaf tissue, which should be dark-adapted for 30 min, cut into strips 1 cm wide or discs of 1 cm diameter and then immersed in 50 μM DCMU for a further 30 min prior to measurement of the fluorescence kinetics. The 50 μM DCMU solution should be prepared by dissolving DCMU in ethanol to a concentration of 1.5 mM and then diluting this solution with water to give a final 50 μM DCMU concentration. Gentle abrasion of the leaf cuticles with a fine nylon brush will enhance the penetration of DCMU into the leaf tissue.

On excitation of dark-adapted, DCMU-treated photosynthetic systems a characteristic fluorescence induction curve is observed (*Figure 22*). Immediately upon excitation the fluorescence rises to a level, designated the minimal fluorescence level and termed F_o. This is the fluorescence level attained when Q_A is maximally oxidised and P = 1.0 in Equation 8 (Section 2.3). The fluorescence yield at F_o, ϕ_{Fo}, is given by:

$$\phi_{Fo} = k_f/[k_f + k_h + k_{t(II \rightarrow I)} + k_p] \quad \text{Equation 17}$$

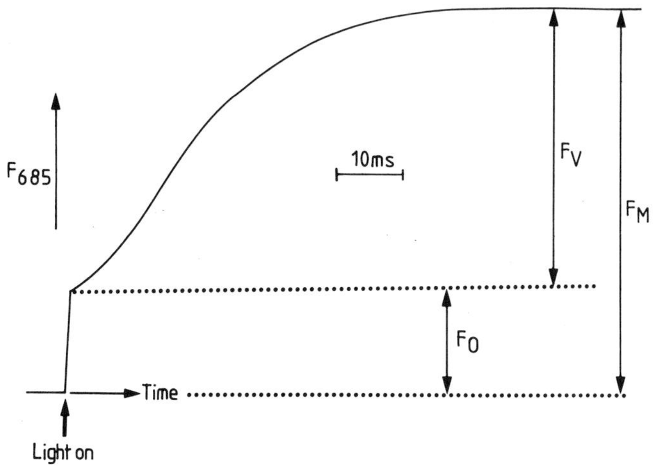

Figure 22. Kinetics of 685 nm fluorescence (F685) induction in dark-adapted, isolated, broken pea chloroplasts in the presence of 15 μM DCMU on excitation with a photon flux density of 100 μmol m^{-2} sec^{-1} of 440 nm radiation. The resuspending medium and excitation source are as described in the legend of *Figure 16*.

Refer to Section 2.3 for the definition of terms. The time taken to reach F_o is determined by the time of opening of the shutter being used. After F_o the fluorescence rises more slowly to a maximal level, termed F_m. The rate of this rise is directly dependent upon the photon flux density used to excite the sample but the relationship breaks down when the pigment beds become light-saturated. At F_m, Q_A is maximally reduced and $P = 0$ in Equation 8 (Section 2.3), thus the fluorescence yield at F_m, ϕ_{Fm}, can be defined by:

$$\phi_{Fm} = k_f/[k_f + k_h + k_{t(II \rightarrow I)}] \qquad \text{Equation 18}$$

The difference in fluorescence at F_o and F_m defines the variable fluorescence, F_v, of the system. The quantum yield of variable fluorescence, ϕ_{Fv}, is thus given by:

$$\phi_{Fv} = \phi_{Fm} - \phi_{Fo} \qquad \text{Equation 19}$$

The ratio of F_v/F_m is a particularly useful parameter since it relates to the quantum yield of photosystem II primary photochemistry, that is, the reduction of Q_A, although it must be emphasised that F_v/F_m does not simply equate to this parameter. F_v/F_m is determined by the product of the quantum yield of photosystem II photochemistry and a term reflecting radiationless de-excitation at the reaction centre of photosystem II (16–18).

Many studies of fluorescence kinetics require accurate measurement of F_o. Good resolution of F_o can be achieved by using a shutter which opens fully in less than 5 msec and a data sampling rate which resolves clearly the transition from the end of the shutter opening to the photochemically-induced rise in the variable fluorescence. Using an excitation photon flux density of 100 μmol m^{-2} sec^{-1} of broad-band blue light with an electromechanical shutter, and a chloroplast sample containing 10 μg chlorophyll ml^{-1}, a data sampling rate of 10 kHz (i.e. 1 data point every 100 μsec) provides a satisfactory resolution of F_o (*Figure 23*). Consistent estimation of F_o is made by measuring F_o at the point at which the extrapolated lines of the initial variable fluorescence

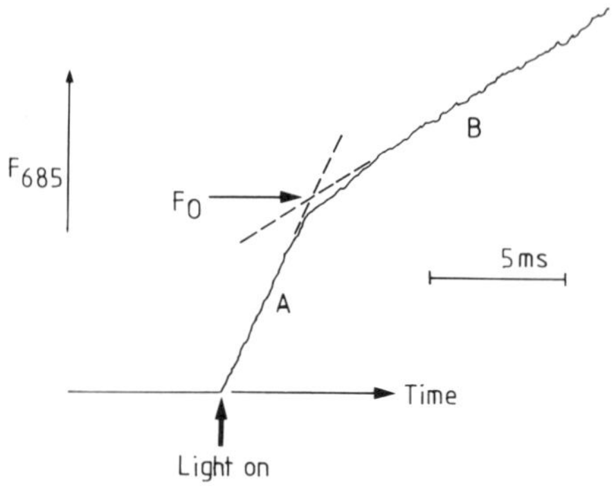

Figure 23. Estimation of the minimal fluorescence level, F_o, from the initial phases of the fluorescence induction curve shown in *Figure 22*. Phase A of the induction represents opening of the shutter, whilst phase B is the beginning of the rise in variable fluorescence. F_o is routinely determined by the intercept of the extrapolation of these two phases.

rise and the rise to F_o meet (*Figure 23*); such extrapolations can be satisfactorily and rapidly done by hand, although microprocessors are frequently employed for the determination. With the experimental system outlined above F_m will be reached in less than 100 msec, thus the commonly used 1024 data point memory sampling at 10 kHz will be sufficient to capture F_m. If experiments are performed with a reduced excitation photon flux density or an increased chlorophyll concentration then the data sampling rate will have to be reduced to accommodate measurement of F_m. In such cases it is advisable to use two transient recorder memories for capture of the fluorescence transients, one sampling at a suitable rate to resolve F_m and the other sampling at 10 kHz (or faster) to resolve F_o.

The kinetics of the variable fluorescence rise from isolated chloroplasts and leaves in the presence of DCMU at 20°C are complex with an initial rapid sigmoid phase being followed by a slower exponential phase (19,20). These biphasic characteristics of the induction curve can be resolved from an analysis of the area growth above the fluorescence curve with time. The reasons for observing such biphasic kinetics have not yet been unequivocally resolved and further discussion of this topic is beyond the scope of this chapter.

At 77 K the fluorescence induction curve is triphasic rather than biphasic. The triphasic nature at 77 K has been attributed to the existence of three electron donors to P680 at cryogenic temperatures; none of these three donors appears to function at 20°C (21). Water cannot act as an electron donor to PSII at 77 K. Since the process limiting the rate of induction of fluorescence at 77 K is the rate of electron donation to P680, and not the rate of reduction of Q_A as is the case at 20°C, analyses of the kinetics of the fluorescence induction cannot provide information on the nature of the PSII electron acceptors. However, the F_o and F_m levels at 77 K represent the fluorescence levels at which Q_A is maximally oxidised and reduced, respectively, thus the definitions of ϕ_{Fo}, ϕ_{Fm} and ϕ_{Fv}, given above for fluorescence at 20°C in the presence of DCMU (equations 17–19), also hold at 77 K.

(ii) *Kautsky curves.* Illumination of dark-adapted leaf tissue or algal cells at ambient temperatures produces a fluorescence induction curve with a number of characteristic

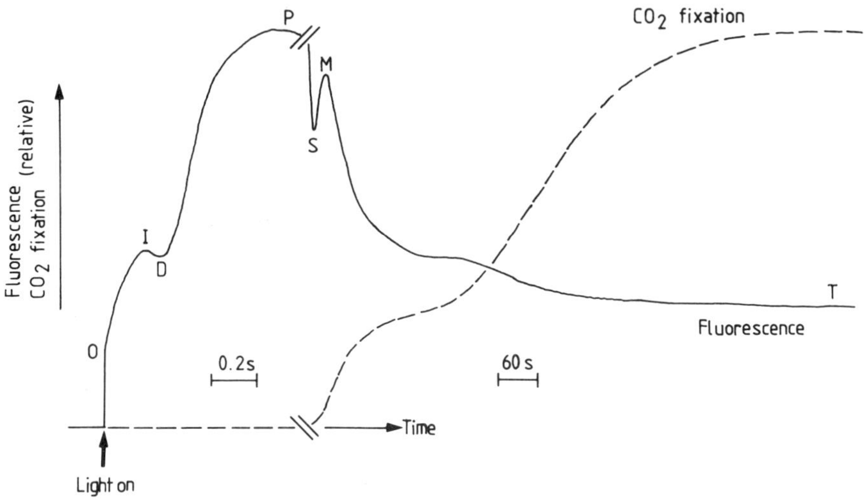

Figure 24. Kinetics of 680 nm fluorescence emission and CO_2 fixation measured simultaneously on exposure of a dark-adapted maize leaf to a photon flux density of 100 μmol m^{-2} sec^{-1} of broad band blue radiation, produced from a quartz-iodine light source and a short pass optical filter with a cut-off at 598 nm. The atmosphere surrounding the leaf contained 350 μmol CO_2 mol^{-1} and 20% O_2 at 23°C. The rate of CO_2 fixation achieved by the leaf at steady state (point T on the induction curve) was 29 μmol m^{-2} sec^{-1}. Note the different time bases before and after point P on the induction curve.

phases which are generally defined by the terminology O, I, D, P, S, M and T for the specific fluorescence levels observed during fluorescence induction (*Figure 24*). It should be noted that modifications in both the shape and duration of the Kautsky curve from a leaf can occur with changes in light intensity. Differences in the characteristics of the curve can also be observed between leaves of the same species but of different ages and from different positions in the canopy and also between leaves of different species. O represents the fluorescence level when Q_A is maximally oxidised. With excitation the fluorescence rises to an 'inflection' point, I, and is then followed by a 'dip' at D before the fluorescence eventually reaches its 'peak' at P. Not all leaf tissues and cells exhibit a clear decrease in fluorescence from I to D; often this transition is represented by a shoulder during the fluorescence induction from O to P. These phases between O and P of the fluorescence induction curve occur rapidly (generally within 2 sec), however the actual rate at which the fluorescence changes occur will depend upon the sample and excitation photon flux density being used. The practical details for the measurement of the OIDP transients are similar to those discussed in the previous section. The rise from O to I is considered to be due to reduction of Q_A, whilst the fluorescence decrease from I to D is indicative of the oxidation of Q_A on electron transfer from Q_A to Q_B. The reduction of Q_A that occurs as the plastoquinone pool becomes increasingly reduced generates the fluorescence rise from D to the maximum at P. At P the plastoquinone pool and Q_A are highly reduced. However, it should be emphasised that in the majority of photosynthetic systems, even under light levels saturating for photosynthesis, Q_A is not maximally reduced at P; the P level of the fluorescence induction is invariably lower than the maximal level of fluorescence, F_m, generated by addition of DCMU to the sample. Q_A remains partially oxidised at

P as a result of electron flow from Q_A to photosystem I.

The kinetics of the slow fluorescence decline from P, through points S and M, to the terminal, stationary fluorescence level at T (*Figure 24*) can be monitored satisfactorily using a good quality potentiometric pen recorder. The PSMT transients in leaf tissue occur over minutes and are related to the induction of CO_2 assimilation in the tissue (22); the appearance of net CO_2 assimilation corresponds with the onset of the S-M transient and thereafter the kinetics of the induction of CO_2 assimilation change in an anti-parallel fashion with the fluorescence emission until both parameters simultaneously reach steady-state at T on the fluorescence induction curve (*Figure 24*). Although many factors can modify the fluorescence yield of thylakoid membranes (23), only two factors are considered to make major contributions to the fluorescence level during P-T quenching in physiologically active, intact photosynthetic systems, such as leaves, algal cells, protoplasts and isolated intact chloroplasts (24,25). The factors are firstly the redox state of Q_A, and secondly, the magnitude of the proton electrochemical potential gradients that exist across the thylakoid membranes. Changes in the redox state of Q_A occur as a result of changes in the rate of non-cyclic electron transport; increased consumption of NADPH by carbon metabolism will result in an increased rate of non-cyclic electron transport and an oxidation of Q_A. Fluorescence quenching due to oxidation of Q_A is termed photochemical quenching and is often abbreviated to q(P) or q_Q in the literature. Changes in the magnitude of the transthylakoid proton electrochemical potential gradients can be related to (i) changes in the rate of proton pumping across the membrane which occur due to changes in electron transport rate and (ii) changes in the rate of utilisation of ATP since ATP synthesis is driven by the transthylakoid proton electrochemical potential gradient (see Chapter 6). Increasing the magnitude of the transthylakoid proton gradient quenches fluorescence from the thylakoids by an as yet unknown physicochemical mechanism. This type of fluorescence quenching is generally referred to as a non-photochemical quenching and is often abbreviated to q_e. It should be stressed that other factors, such as phosphorylation of LHCII, depletion of cation levels around the thylakoids and generation of reduced phaeophytin, will also quench fluorescence in a non-photochemical fashion; however under non-stress conditions in fully physiologically competent photosynthetic systems, alteration of the transthylakoid proton electrochemical potential gradient is by far the most significant factor contributing to changes in non-photochemical quenching.

7.3.3 *Advanced Techniques for Analysis of Fluorescence Induction*

The previous section has considered the theoretical basis and practical measurement of the Kautsky curve from photosynthetic organisms. The following sections present more advanced methods for examining the Kautsky phenomena, which in the main require more advanced instrumentation.

(i) *Estimation of photochemical and non-photochemical fluorescence quenching.* The rapid addition of DCMU to a stirred solution of isolated thylakoids, intact chloroplasts, protoplasts and algal cells at steady-state fluorescence at T produces a biphasic reversion of fluorescence quenching (*Figure 25*). It is essential to add the DCMU rapidly and in the dark by using a hypodermic syringe and ensure rapid mixing of the DCMU

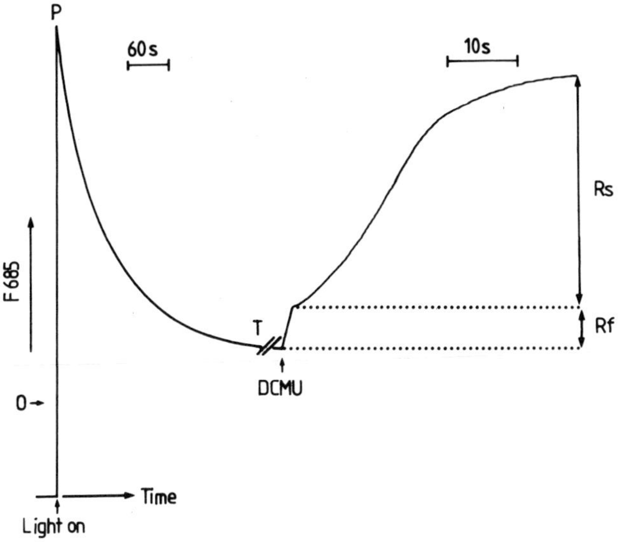

Figure 25. Effect of addition of 10 μM DCMU on 685 nm fluorescence from isolated, intact pea chloroplasts at the steady-state fluorescence level (point T). R_f and R_s represent the fast and slow phases of the reversal of fluorescence, respectively, on addition of DCMU.

with the sample solution by rapidly stirring the sample throughout the DCMU addition. DCMU should be added to a final concentration of $10-15$ μM. The rapid phase of the DCMU-induced fluorescence increase, R_f, has a half-rise time of approximately 1 sec and is the result of a rapid reduction of Q_A. The fluorescence level attained at the end of this phase represents the fluorescence level that would be observed from the experimental system if Q_A were maximally reduced, a level designated $F_{(Qred)}$. R_f represents the amount of fluorescence at T that was quenched by oxidation of Q_A. The slow phase of the DCMU-induced fluorescence increase, R_s, can have a half-rise time ranging from about 5 to 15 sec depending upon the photosynthetic system being examined. R_s results from the relaxation of the proton electrochemical potential gradient across the thylakoid membranes. Measurement of R_f and R_s allows estimation of the amounts of fluorescence quenching that have occurred in the experimental system due to photochemical and non-photochemical processes. This technique has enabled resolution to be made of the amounts of fluorescence quenching attributable to photochemical processes and to the generation of a transthylakoid proton electrochemical potential gradient during the SMT transients in algal cells, isolated and intact chloroplasts (24,25) and protoplasts (26). Unfortunately, this DCMU-addition technique can only be used with photosynthetic systems in solution; it cannot be used with leaf tissue due to the slow rate of penetration of the DCMU to the thylakoids within the leaf cells.

The redox state of Q_A in leaf tissue can be examined by analysis of the fluorescence transients generated by the addition of a second, saturating irradiation of the leaf tissue (27,28). On addition of the second irradiation to the leaf an instantaneous rise in fluorescence emission to a level termed F_{O2} (*Figure 26*) is seen. If prior to addition of the second excitation Q_A was maximally reduced, then the fluorescence would simply rise to F_{O2} and remain at this level in the short term. If Q_A was partially oxidised the ad-

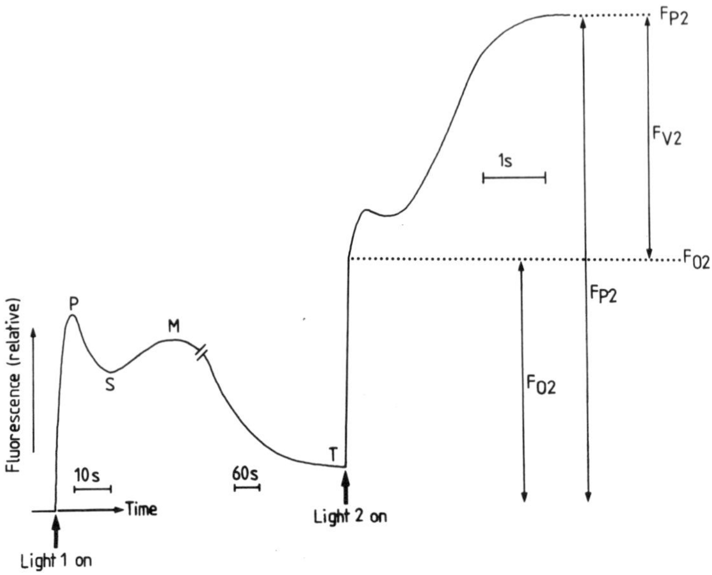

Figure 26. The kinetics of 685 nm fluorescence emission from a dark-adapted bean leaf on exposure initially to a photon flux density of 100 μmol m^{-2} sec^{-1} of 633 nm radiation (Light 1) and then on exposure to a further 500 μmol m^{-2} sec^{-1} of 633 nm radiation (Light 2) on reaching a steady state fluorescence level at T. The 633 nm radiation was produced from a 10 mW helium-neon laser in conjunction with a 633 nm interference filter. Note the difference in the time bases used before and after the addition of the second light at T.

dition of the second excitation would generate some variable fluorescence, designated F_{V2}, after F_{O2} had been reached (*Figure 26*). The magnitude of the ratio of F_{V2}/F_{O2} is related to the proportion of Q_A oxidised prior to addition of the second excitation (27,28). The fluorescence level that would be obtained from the leaf tissue immediately prior to excitation with the second irradiation if Q_A were maximally reduced, that is, $F_{(Qred)}$, can be estimated from:

$$F_{(Qred)} = F_{(add)} + [F_{(add)} \times (F_{V2}/F_{O2})] \qquad \text{Equation 20}$$

where $F_{(add)}$ is the fluorescence level of the tissue prior to addition of the second irradiation (*Figure 26*). $F_{(Qred)}$ can be estimated using this procedure throughout a fluorescence induction curve and the difference between $F_{(Qred)}$ and the observed fluorescence level gives the amount of fluorescence that has been quenched by photochemical processes (*Figure 27*). When measuring the fluorescence induction curve generated by the saturating second irradiation, it is essential to use a shutter which opens in less than 3 msec, since if Q_A is highly reduced the maximal fluorescence level at F_{P2} (see *Figure 26*) can be reached within 20 msec, thus making resolution of F_{O2} and F_{V2} difficult if a rapidly opening shutter is not used. A data sampling frequency of the order of 10–20 kHz should be used to determine F_{O2}, whilst a reduced frequency can often be used for measurement of F_{V2}. The intensity of the second irradiation should be carefully chosen to produce maximal F_{V2}/F_{O2} values. For bean leaves a photon flux density of 500 μmol m^{-2} sec^{-1} was saturating; however, at greater intensities F_{V2}/F_{O2} decreased due to photoinhibitory damage to the photosynthetic apparatus resulting in a quenching of F_{V2} but not of F_{O2} (28). The second irradiation

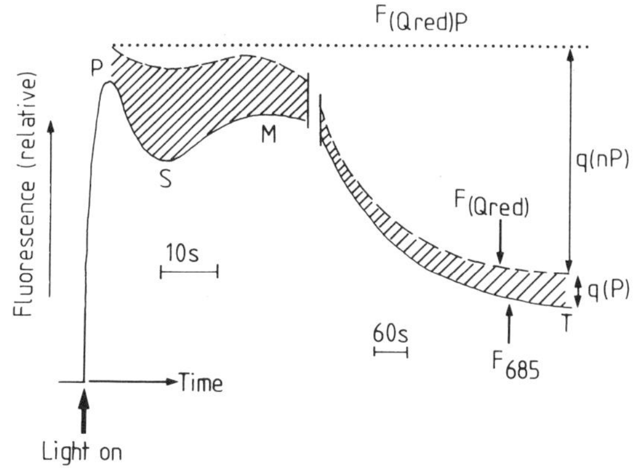

Figure 27. Photochemical and non-photochemical components of fluorescence quenching during the P to T transition in bean leaves. The solid line represents the 685 nm fluorescence (F685) induction curve. The level of fluorescence that would occur from maximal Q_A reduction, i.e. $F_{(Qred)}$, is given by the dashed line. Fluorescence quenching from $F_{(Qred)}$ at point P (i.e. $F_{(Qred)P}$, which is given by the dotted line) to $F_{(Qred)}$ at any given point on the induction curve is the result of non-photochemical quenching, q(nP), processes. Fluorescence quenching from the $F_{(Qred)}$ level to the actual observed fluorescence level, i.e. F685, on the induction curve is the result of photochemical, q(P), processes (i.e. shaded area).

must excite exactly the same area of leaf tissue as the primary excitation otherwise the relationship between F_{V2}/F_{O2} and the redox state of Q_A will not hold. This can be achieved by using a mask over the leaf tissue which has an aperture slightly smaller than the smallest area of tissue excitation provided by the two light sources. Alternatively, a trifurcated fibre optic with the fibres randomised in the common end can be used. The two excitation beams are passed down two arms of the fibre optic and reach the leaf *via* the randomised common end of the fibre optic, whilst the third arm is used to transmit fluorescence emission to a detector.

A modification of the light-addition method described above for estimating the amount of photochemical quenching in leaf tissue which employs a modulated light source to generate chlorophyll fluorescence and drive photosynthesis, has recently been developed (26). Using a lock-in amplifier to monitor the modulated fluorescence signal, it is possible to expose the leaf to high intensity light from a continuous light source, which will maximally reduce Q_A, and monitor directly the rise in modulated fluorescence due to the reduction of Q_A. This method removes the requirement for measurement of F_{O2} (see *Figure 26*), since the lock-in amplifier detection system monitors only the rise in the fluorescence signal generated by the modulated light source, not the fluorescence signal produced by the continuous high-intensity light. The principles of using modulated fluorescence measurements are discussed below.

(ii) *Modulated fluorescence.* One of the major drawbacks with the measurement of continuous fluorescence signals from a photosynthetic system is that the sample cannot be excited with radiation containing any of the wavelengths of fluorescence emission being studied. This difficulty precludes the use of white light and far-red light (the latter preferentially excites photosystem I) in photosynthetic investigations involving measurements of continuous fluorescence. By using a modulated light source to excite

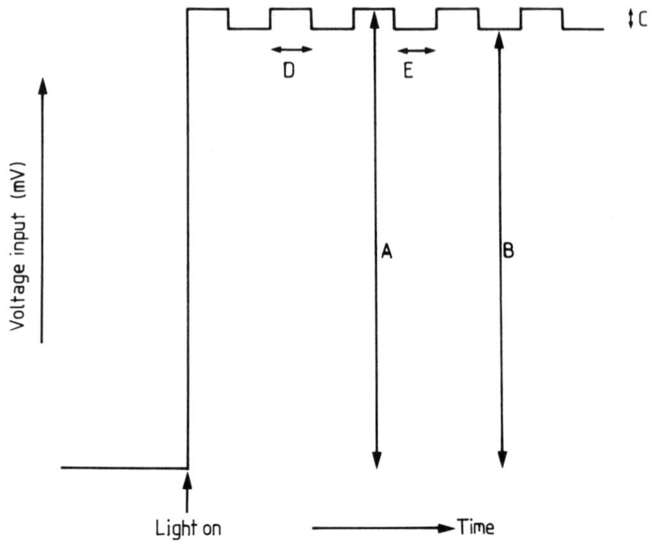

Figure 28. Voltage input to the lock-in amplifier from a photodetector which is monitoring the 685 nm radiation received from a photosynthetic system exposed simultaneously to modulated low intensity, broad band blue or yellow radiation and high intensity white actinic radiation. On exposure to both the modulated and actinic light sources (e.g. during period D) the electrical signal reaches a level A, which is due to the fluorescence generated from the sample by the two light sources and the actinic light reflected from the sample surface. When the modulated light source is off and the actinic light is on (e.g. during period E), a signal level B is reached, which is due to the fluorescence generated from the sample by the actinic source and the actinic light reflected from the sample surface. When the modulated light source is off and the actinic light is on (e.g. during period E), a signal level B is reached, which is due to the fluorescence generated from the sample by the actinic source and the actinic light reflected from the sample surface. The difference between the signal levels A and B gives the signal due to fluorescence generated by the modulated source, i.e. C. C is the signal that is measured and amplified by the lock-in amplifier. D is the period over which the modulated light is on. E is the period during which the modulated light is off.

fluorescence in conjunction with a fluorescence detection system which monitors only fluorescence emitted at the frequency of the modulated exciting light, it is possible to excite the sample with actinic d.c. light containing the same wavelengths as the monitored fluorescence and yet monitor only the fluorescence generated by the modulated light source. The d.c. fluorescence signal produced by excitation of the sample with the actinic d.c. light source is not monitored by the detection system. The detection system can consist of either a photodiode or photomultiplier connected to a lock-in amplifier, which is ‘tuned’ to the frequency of the modulated excitation light. The lock-in amplifier monitors the difference between the signals generated by the photodetector on receiving (i) the reflected and fluoresced continuous radiation from the sample (level B in *Figure 28*) and (ii) the reflected and fluoresced continuous radiation and the modulated fluorescence from the sample (level A in *Figure 28*). A d.c. light source, used in conjunction with a mechanical or electrical beam-chopping device, can provide a suitable modulated light source. Alternatively, an electrical pulse generator can be used to modulate a light source; light-emitting photodiodes are particularly suitable as electrically modulated light sources, although their maximum light emission may be rather low for many photosynthetic experiments. It is preferable to have the frequency of light

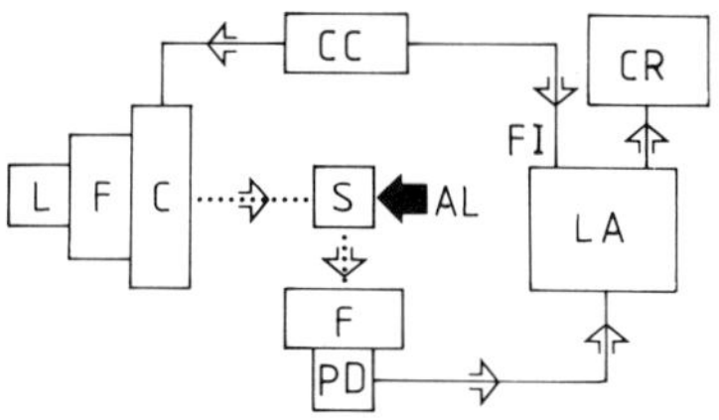

Figure 29. Apparatus for the generation and measurement of modulated fluorescence from a photosynthetic sample in solution. AL, actinic light; C, chopper; CC, chopper control unit; CR, chart recorder; F, optical filter; FI, frequency input; L, light source; LA, lock-in amplifier; PD, photodetector; S, sample.

modulation above 100 Hz and not use multiples of the frequency of the mains electrical supply (e.g. 50 Hz in the UK and 60 Hz in the USA) to prevent harmonic interference from mains-driven electrical equipment. A block diagram of an apparatus suitable for the generation and measurement of modulated fluorescence is shown in *Figure 29*. Practical difficulties can be experienced when first using a lock-in amplifier, consideration of which are beyond the scope of this chapter, and it is recommended that when attempting to set up a modulated fluorescence detection system professional advice should be sought. It seems likely that in the near future instruments for the measurement of modulated fluorescence will become commercially available. The use of a commercial instrument will circumvent the bulk of the technical difficulties experienced when setting up a modulated detection system from scratch.

In many applications of the modulated fluorescence technique in the study of photosynthesis, it is useful to use a modulated light intensity sufficiently low that it does not generate any variable fluorescence from the sample. The level of fluorescence generated by the modulated light in this case is analogous to the minimal fluorescence level, F_0 (see *Figure 22*), that is generated immediately on exposure to continuous light. Using broad-band yellow modulated light, produced from light-emitting photodiodes pulsed at 135 Hz, a photon flux density of 1.15 μmol m^{-2} sec^{-1} at the leaf surface has been found to be sufficiently low to maintain the modulated fluorescence signal at the F_0 level (29). If different excitation wavelengths which are more efficiently absorbed by the sample (e.g. blue or red light) are used then lower photon flux densities will be required to prevent generation of any variable fluorescence. Exposure of the sample, whilst being irradiated with the weak modulated light, to actinic light will produce a variable fluorescence emission, which will be observed as a variable component of the modulated fluorescence signal. Thus using a weak modulated light source to generate modulated fluorescence, the effects of actinic light (which can contain excitation wavelengths the same as the monitored, modulated fluorescence emission wavelength) on the modulated fluorescence characteristics of the thylakoids can be studied. In such studies with actinic light it is essential to ensure that the photodiode used to detect the modulated fluorescence does not become saturated as a result of reflection of the actinic light into the photodiode. An example of the use of modulated fluorescence generated by a weak modulated light source on exposure of a leaf to high intensity white light is shown in *Figure 30*. On excitation of the leaf with the weak modulated light the fluorescence rises to a level analogous to F_0; no variable fluorescence is generated. On exposing the leaf to 200 μmol m^{-2} sec^{-1} of white ac-

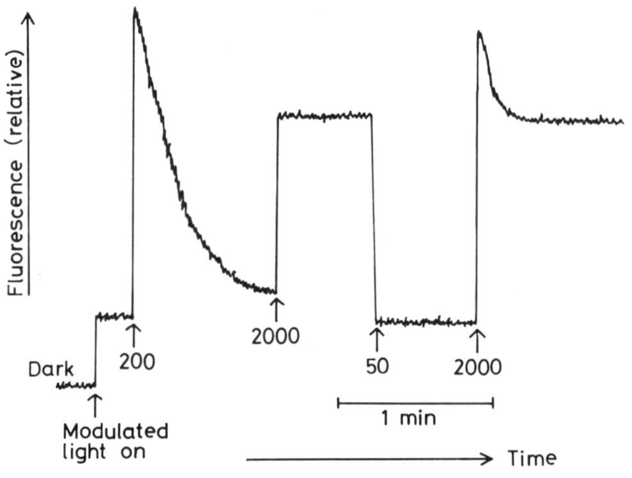

Figure 30. Changes in modulated fluorescence emission from a navel orange leaf exposed to a photon flux density of 1.15 μmol m^{-2} sec^{-1} of modulated yellow light and then exposed to different photon flux densities of white actinic light. Arrows show the times at which the actinic light intensity was changed. The photon flux density of the actinic light is given in μmol m^{-2} sec^{-1} below each arrow.

tinic light modulated fluorescence rapidly rises and then slowly declines. The rapid rise is attributed to rapid reduction of Q_A, whilst the slow decline is caused by re-oxidation of Q_A and an increase in the magnitude of the proton electrochemical potential gradients across the thylakoids. On increasing the actinic light from 200 to 2000 μmol m^{-1} sec^{-1} the modulated fluorescence rises, as a result of Q_A reduction, to a steady-state level that is lower than the maximum attained with 200 μmol m^{-2} sec^{-1} of white light. This reduced maximal fluorescence level with a 10-fold increase in actinic light intensity is probably a demonstration of the existence of a considerably greater transthylakoid proton electrochemical potential gradient prior to exposure to 2000 μmol m^{-2} sec^{-1} of white light, than existed in the weak modulated light (29). The absence of fluorescence quenching in 2000 μmol m^{-2} sec^{-1} of white light suggests that neither re-oxidation of Q_A nor increased thylakoid energisation occurs on prolonged exposure to 2000 μmol m^{-2} sec^{-1}. Reduction of the actinic white light from 2000 to 50 μmol m^{-2} sec^{-1} rapidly reduces the modulated fluorescence level. On returning the leaf to 2000 μmol m^{-2} sec^{-1} of white actinic light, the modulated fluorescence rises to a maximum, which is greater than the previous maximum achieved with this light intensity, indicating the existence of a reduced transthylakoid proton electrochemical potential gradient at steady-state in 50 compared with 200 μmol m^{-2} sec^{-1} of actinic light, since Q_A would be expected to be similarly reduced at both of these two maxima. This hypothesis is supported by the quenching of the modulated fluorescence to a level similar to the steady-state, with the first exposure to 2000 μmol m^{-2} sec^{-1} after the final maximum in the 2000 μmol m^{-2} sec^{-1} light level has been reached.

(iii) *Fluorescence decay kinetics.* Fluorescence decay kinetics (fluorescence lifetimes) can provide information on the origin of particular fluorescence signals from the photosynthetic membrane, and on excitation energy transfer between pigment beds (30). However, measurement of fluorescence decay kinetics requires highly specialised, expensive equipment that is available in only a limited number of laboratories. Since detailed discussion of this topic is beyond the scope of this chapter, it is raised here only to make the reader aware of the technique and enable the literature on fluorescence decay kinetics to be considered in the context of the reader's own experiments. In studies of fluorescence decay kinetics samples are usually excited by a pulsed mode-locked laser that can provide trains of short duration (10 – 15 psec) flashes. Generally, streak cameras are used to resolve the components of the kinetics of fluorescence emission, since the lifetimes of the kinetic components of fluorescence emission from chloroplasts are extremely short, ranging from about 40 psec to about 2.5 nsec (30). Alternatively, very rapidly modulated continuous excitation light is used and the phase shift between the actinic light and the fluorescence emission is measured (31).

8. OTHER SPECTROSCOPIC TECHNIQUES

This section outlines the principles of spectroscopic techniques other than absorbance and fluorescence spectroscopy that are used in photosynthesis research. The techniques require sophisticated and specialised equipment, so the aim of this section is to give a brief outline of how the techniques work, rather than give details of procedures. The interested reader can find a more detailed treatment in Brown (32).

8.1 **Circular Dichroism**

It is well known that a molecule containing a single asymmetric carbon atom is optically active, and that if plane polarised light is passed through a solution of such molecules, then the plane of polarisation is rotated. Circular dichroism (c.d.) is an extension of this technique, and employs circularly polarised light as the measuring beam. In a c.d. spectrometer, the output signal is the difference between the absorption of left handed and right handed circularly polarised light, which are alternately shone onto the sample.

The interpretation of c.d. spectra is complex, but a c.d. signal indicates molecular asymmetries. For example, monomeric chlorophyll has a very weak c.d. spectrum, but chlorophyll-chlorophyll interactions, or more importantly, chlorophyll-protein interactions can give strong c.d. signals. A typical application of c.d. is to check the interactions between pigment and protein in complexes *in vivo* and *in vitro*. If the c.d. spectrum of an isolated complex resembles that of the complex *in vivo*, then there is an indication that the complex has not drastically been altered during the isolation.

8.2 **Electron Paramagnetic Resonance Spectroscopy**

Electron paramagnetic resonance (e.p.r.; also known as electron spin resonance, e.s.r.) is a technique for detecting unpaired electrons. When placed in an applied magnetic field, the spin of an unpaired electron can assume one of two orientations with respect to the field; one will be of lower energy, and the other of higher energy. A transition between the two orientations can occur when a photon of the correct energy is absorbed, in an analogous way to the electronic transitions discussed in Section 2.1. In an e.p.r. spectrometer the sample is subjected to a microwave field ($\sim$10 GHz) and is held in a magnetic field whose field strength is gradually scanned across the region of interest. The spectrometer measures the absorption of the microwave field; at certain values of the applied magnetic field there is a 'resonance' or strong absorption, when

$$h\nu = \Delta E = g \beta B_0$$

where $h\nu$ is the energy of the microwave photon, ΔE is the energy difference between the two orientations of the electron spin with respect to the applied magnetic field, g is a dimensionless number, β is the Bohr magneton and B_0 is the applied magnetic field strength. An e.p.r. spectrum is a plot of microwave absorption against applied magnetic field strength, but the first derivative is usually presented. The position of a resonance is determined by the value of g: for a free electron g = 2.0023 and the resonance has a symmetrical shape, whilst for transition metal ions, the value of g can be significantly removed from 2, and the spectrum can show considerable anisotropy.

In photosynthesis, e.p.r. has been applied to unpaired electrons in the non-haem iron-sulphur (Fe-S) centres, the reaction centre chlorophylls of photosystems I and II, quinones in the electron transfer chain, and manganese in the oxygen evolving complex (33). In some cases special e.p.r. spectrometers have been used where the sample can be illuminated whilst it is in the instrument. For some samples, especially Fe-S centres, spectra are gathered with the sample at low temperature (liquid nitrogen or liquid helium).

The technique is useful in cases where optical spectroscopy fails to detect redox changes: e.p.r. transitions are characterised by their g values and by the degree to which the transition is broadened. Since some measurements are made at low temperature, the technique lends itself to the study of the most primary oxidation and reduction events, and has often been used in conjunction with redox potentiometry.

8.3 Nuclear Magnetic Resonance

Nuclear magnetic resonance (n.m.r.) spectroscopy is in many ways analagous to e.p.r., except that the spin of the nucleus is considered. A detailed description of the technique is beyond the scope of this chapter, particularly as n.m.r. has not yet been applied extensively to photosynthetic systems. A comprehensive treatment is given in Brown (32).

The technique is not limited to paramagnetic species, but for biological studies it turns out that a number of nuclei is suitable for study by n.m.r., including ^{1}H, ^{13}C, ^{19}F, and ^{31}P. Nuclear magnetic resonance has been widely used by chemists for gathering structural information about small molecules; more recently n.m.r. has been applied to biological molecules and systems.

In photosynthesis, n.m.r. has been used to try to probe the release of protons by the water-splitting complex of photosystem II, and has also been applied to the study of the levels of phosphorylated intermediates of the Calvin cycle.

9. ACKNOWLEDGEMENTS

We thank Mr A. Anderson and Dr M. Bradbury for assistance with preparing the figures.

10. REFERENCES

1. Ludlow,M. (1982) in *Techniques in Bioproductivity and Photosynthesis,* Coombs,J. and Hall,D.O. (eds.), Pergamon Press, Oxford, p. 5.

2. Clayton,R.K. (1970) *Light and Living Matter, Vol 1, The Physical Part*, published by McGraw Hill, New York.
3. Holden,M. (1976) in *Chemistry and Biochemistry of Plant Pigments*, Vol. **2**, 2nd edition, Goodwin,T.W. (ed.), Academic Press, London, p. 1.
4. Mackinney,G. (1941) *J. Biol. Chem.*, **140**, 315.
5. Arnon,D.I. (1949) *Plant Physiol.*, **24**, 1.
6. Shibata,K. (1959) in *Methods of Biochemical Analysis*, Vol **7**, Glick,D. (ed.), Interscience John Wiley, New York, p. 77.
7. Butler,W.L. (1964) *Annu. Rev. Plant Physiol.*, **15**, 451.
8. Butler,W.L. (1972) in *Methods in Enzymology*, Vol **24**, San Pietro,A. (ed.), Academic Press, London, p. 3.
9. Bendall,D.S., Davenport,H.E. and Hill,R. (1971) in *Methods in Enzymology*, Vol **23**, San Pietro,A. (ed.), Academic Press, London, p. 327.
10. Marsho,T.V. and Kok,B. (1980) in *Methods in Enzymology*, Vol **69**, San Pietro,A. (ed.), Academic Press, London, p. 280.
11. Lodola,A. (1984) in *Microcomputers in Biology – A Practical Approach*, Ireland,C.R. and Long,S.P. (eds.), IRL Press, Oxford and Washington, D.C., p. 179.
12. Skoog,D.A. (1985) *Principles of Instrumental Analysis*, CBS College Publishing, New York.
13. Samikhatova,O.A., Chulanovskaya,M.V. and Metzner,H. (1971) in *Plant Photosynthetic Production. Manual of Methods*, Sestak,Z., Catsky,J. and Jarvis,P.G. (eds.), Dr. W. Junk N.V. Publishers, The Hague, p. 238.
14. Bose,S. (1982) *Photochem. Photobiol.*, **36**, 725.
15. Krause,G.H., Briantais,J.M. and Vernotte,C. (1983) *Biochim. Biophys. Acta*, **723**, 169.
16. Butler,W.L. and Kitajima,M. (1975) *Biochim. Biophys. Acta*, **396**, 72.
17. Van Grondelle,R. and Duysens,L.N.M. (1980) *Plant Physiol.*, **65**, 751.
18. Haehnel,W., Nairn,J.A., Reisberg,P. and Sauer,K. (1982) *Biochim. Biophys. Acta*, **680**, 161.
19. Melis,A. and Homann,P.H. (1975) *Photochem. Photobiol.*, **21**, 431.
20. Hipkins,M.F. (1978) *Biochim. Biophys. Acta*, **502**, 514.
21. Bonnet,F., Vernotte,C., Briantais,J.-M. and Etienne,A.-L. (1977) *Biochim. Biophys. Acta*, **461**, 151.
22. Ireland,C.R., Long,S.P. and Baker,N.R. (1984) *Planta*, **160**, 550.
23. Krause,G.H. and Weis,E. (1984) *Photosynth. Res.*, **5**, 139.
24. Krause,G.H., Briantais,J.-M. and Vernotte,C. (1982) *Biochim. Biophys. Acta*, **679**, 116.
25. Horton,P. (1983) *Proc. R. Soc. Lond. Ser. B*, **217**, 405.
26. Quick,P. and Horton,P. (1984) *Proc. R. Soc. Lond. Ser. B*, **220**, 371.
27. Bradbury,M. and Baker,N.R. (1981) *Biochim. Biophys. Acta*, **635**, 542.
28. Bradbury,M. and Baker,N.R. (1984) *Biochim. Biophys. Acta*, **765**, 275.
29. Ögren,E. and Baker,N.R. (1985) *Plant Cell Environ.*, **8**, 539.
30. Karukstis,K.K. and Sauer,K. (1983) *J. Cell Biochem.*, **23**, 131.
31. Moya,I. and Garcia,R. (1983) *Biochim. Biophys. Acta*, **722**, 480.
32. Brown,S.B., ed. (1980) *An Introduction to Spectroscopy for Biochemists*, published by Academic Press, London.
33. Bearden,A.J. and Malkin,R. (1974) *Q. Rev. Biophys.*, **7**, 131.

CHAPTER 5

Electron Transport and Redox Titration

J.F. ALLEN and N.G. HOLMES

1. INTRODUCTION

Photosynthesis is essentially a light-driven oxidation-reduction reaction involving coupled synthesis of ATP. Even a whole-plant physiologist measuring leaf CO_2 uptake can therefore be said to be engaged in measurement of electron transport by indirect means. This chapter is restricted to more direct, continuous measurement of electron transport through pathways located primarily in the chloroplast thylakoid membrane system, and with redox titration of individual components of such pathways. Measurement of electron transport in intact chloroplasts and in leaf discs is also discussed. Assay of some individual electron carriers is covered in Chapter 4, while ATP synthesis coupled to electron transport is described in Chapter 6.

1.1 **Electron Transfer**

When one or more electrons are transferred from a species A to a species B, A becomes oxidised and B becomes reduced.

$$A_{red} + B_{ox} \xrightarrow{ne^-} A_{ox} + B_{red}$$

In such an oxidation-reduction ('redox') reaction A may be described as the electron donor or reductant and B as the electron acceptor or oxidant: A *reduces* B and B *oxidises* A. The whole transfer may be viewed as the sum of two half-reactions, oxidation of the donor, A

$$A_{red} \rightarrow A_{ox} + ne^-$$

and reduction of the acceptor, B

$$B_{ox} + ne^- \rightarrow B_{red}$$

Any particular half-reaction, or couple, has a specific mid-point redox potential which depends on the identity of the chemical species involved. The direction of electron transfer as any two half-reactions approach equilibrium can be predicted from their individual mid-point redox potentials — the donor will be the species with the half-reaction of lower potential. The mid-point redox potential at pH 7.0 (i.e. the standard mid-point redox potential) is written E_{m7}. These terms are more rigorously defined in Section 3.1. Measurement of E_{m7} for any electron carrier is covered in Section 3.

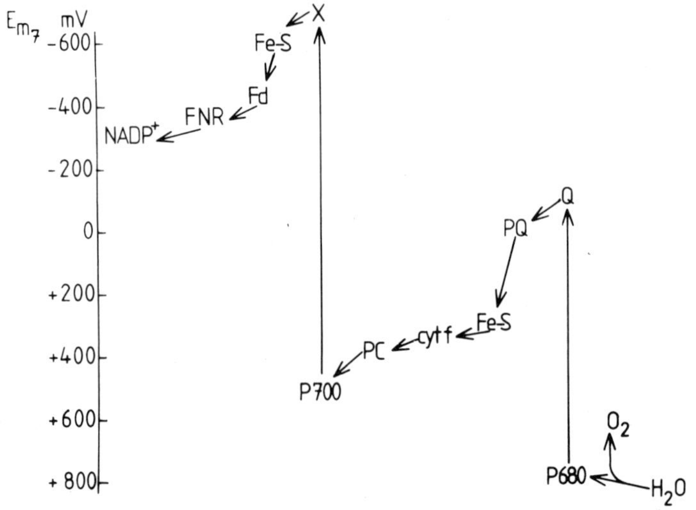

Figure 1. The 'Z-scheme' for non-cyclic electron transport in photosynthesis.

1.2 The Electron Transport Chain of Plant Photosynthesis

The Z-scheme for photosynthetic electron transport (*Figure 1*) is obtained simply by plotting the sequence of electron carriers on a vertical scale of E_{m7}. Two abrupt drops in E_{m7} correspond to the two photochemical reactions of plant-type photosynthesis, and the Z-scheme model when first proposed (1) provided an explanation for a number of lines of evidence suggesting that two such light reactions are linked in series. In fact this model applies only to the electron transport chains of green plants, algae and cyanobacteria; the purple and green photosynthetic bacteria have single photosystems driving cyclic or non-cyclic electron transport. Although this volume is concerned largely with green-plant photosynthesis, bacterial photosynthesis has played an essential role in the development of this subject (2) and comparisons of plants and bacteria are continuing to be highly informative for students and research workers alike (3).

In the Z-scheme model the two photochemical reactions are driven by distinct pigment systems, and are termed photosystem I (PSI, the first to be discovered) and photosystem II (PSII). PSI drives electron transport from plastocyanin (or a *c*-type cytochrome in certain algae) at $E_{m7} = +390$ mV to ferredoxin at $E_{m7} = -420$ mV. PSII drives electron transport from water at $E_{m7} = +820$ mV to plastoquinone at $E_{m7} = 0$ mV. Electron transport from plastoquinone occurs 'downhill' through a cytochrome complex which therefore links PSI and PSII. Electron transport from ferredoxin to $NADP^+$ at $E_{m7} = -320$ mV occurs 'downhill' through a flavoprotein, ferredoxin-NADP reductase. Certain electron carriers associated with each photosystem have even lower E_{m7} values than plastoquinone and ferredoxin. These participate in the primary photochemical reactions and are probably phaeophytin together with certain plastoquinone species (PSII) and chlorophyll together with additional iron-sulphur species (PSI).

The whole electron transport chain therefore requires a minimum of two quanta of radiation (one for each photosystem) in order to drive each electron uphill through a total redox span of 1140 mV. The end products are oxygen, NADPH and ATP. It is also clear that PSI can pass electrons to plastoquinone and hence to itself, thereby driving a cyclic electron transport pathway the end-product of which is ATP alone.

With this minimum of background information on the electron transport chain it should be possible for the reader to make sense of the techniques of measurement to be described. Adoption of a practical approach will not, of course, remove the need to begin by asking the right sort of question.

Familiarity with the field of investigation through reviews (4) and primary research articles is presumably a minimal requirement for the design of informative experiments.

2. MEASUREMENT OF ELECTRON TRANSPORT

2.1 Principle of Measurement: the Hill Reaction

Non-cyclic electron transport from water to any acceptor through all or part of the chain depicted in *Figure 1* can be described in a single equation:

$$H_2O + A \xrightarrow{2e^-} AH_2 + \frac{1}{2}O_2$$

This is one formulation of the Hill reaction, which is the light-dependent evolution of oxygen by isolated chloroplasts. It requires reduction of any suitable added electron acceptor, or 'Hill oxidant', represented by A in the equation.

Measurement of non-cyclic electron transport can therefore take the form of measurement of the oxygen produced, or of reduction of the electron acceptor. Both these principles of measurement are widely used in photosynthesis research, though oxygen evolution is obviously impossible for related reactions where an artificial electron donor replaces water. Measurement of reduction of the acceptor is also preferred where the stoichiometry of net oxygen evolution is in doubt (due perhaps to competing oxygen uptake). With these two exceptions, measurement of oxygen evolution is generally the most convenient and widely used principle of measurement.

2.2 Comparison of Techniques

As indicated above, measurement of oxygen evolution or uptake is frequently the method of choice for determining electron transport rates. Before considering the oxygen electrode in detail (Section 2.3), we briefly review the other techniques which are available for measuring electron transport.

2.2.1 *Manometry*

Before the 1960s, most measurements of electron transport involved manometric techniques of one kind or another, and high sensitivity and accuracy could be obtained by skilled practitioners of the art. On the whole the oxygen electrode (Section 2.3) has replaced the Warburg manometer as the standard method of measuring photosynthetic electron transport, and such special applications of manometry that remain are likely to be displaced by the increasing range and versatility of polarographic sensors. Anyone determined to use manometry is referred to the comprehensive treatise of Umbreit *et al.* (5).

2.2.2 *Spectrometry*

Optical absorbance, fluorescence and electron paramagnetic resonance spectroscopy are all used extensively in photosynthesis research and their major applications are described in Chapter 4. Infra-red gas analysis (IRGA) is a favoured method of measuring carbon dioxide uptake, having replaced the manometric techniques originally employed in this area, though many routine IRGA measurements are now likely to be replaced by the simpler and less expensive leaf-disc oxygen electrode (Section 2.8) or by more direct polarographic measurement of CO_2. Mass spectrometry using [^{18}O]- and [^{16}O]oxygen was important historically in providing decisive evidence for photolysis of water as the source of evolved oxygen, and continues to be invaluable where it is necessary to discriminate between oxygen uptake and oxygen evolution in the same experimental system (6).

A more recently developed technique for measuring photosynthetic electron transport is photoacoustic spectrometry, where modulated light provides pulses of oxygen evolution which may be detected as pressure changes with a microphone. The acoustic signal is amplified by a phase-sensitive detector which is locked to the frequency of the modulated light. Photoacoustic measurements of complex transients (e.g. state 1–state 2 transitions; see Chapter 4) are now possible, and the technique provides a promising, non-invasive way of measuring photosynthetic yield in intact systems (e.g. leaves) directly rather than by having to infer it from (equally non-invasive) fluorescence yield (7). A photoacoustic sensor is available from Applied Photophysics, London (see Appendix to chapter).

It is likely that ion-selective electrodes will become increasingly used in measurement of photosynthetic electron transport. As indirect measurements, both carbon dioxide and nitrate or nitrite electrodes have been used in algae, bacteria and intact chloroplasts and protoplasts. A sulphide electrode has been used for electron transport measurement in green bacteria. The chief advantage of polarographic techniques is that they provide continuous measurement: they are also comparatively cheap and easy to use. It is likely that the new technology of biosensors will produce even more selective and useful polarographic techniques in due course. Meanwhile, the most useful polarographic technique, at least for plant-type photosynthesis, is the oxygen electrode (Section 2.3).

2.2.3 *Absorption Changes of an Added Electron Acceptor on its Reduction in the Hill Reaction*

A suitably buffered chloroplast suspension prepared by the techniques described in Chapter 2 will reduce any one of a range of electron acceptors (Hill oxidants) in the light. A low-cost way of demonstrating the electron transport activity of isolated chloroplasts is to illuminate the chloroplast suspension for a period of time (say 10 min) and to measure an optical absorbance change caused by reduction of the electron acceptor. Suitable controls are obtained by leaving an identical sample in the dark or by illuminating another sample in the presence of the electron transport inhibitor 3-(3,4-dichlorophenyl)-1,1-dimethylurea (DCMU) at a final concentration of at least 10 μM.

One widely used electron acceptor is the dye 2,6-dichlorophenolindophenol (DCPIP or DCIP). The oxidised form of DCPIP has a broad red absorbance band centred at 590 nm and appears blue. On reduction the red absorbance band disappears and the solution becomes colourless — this can be demonstrated by adding the reductant ascorbic

Table I. Properties of Three Electron Acceptors Whose Reduction can be Measured by Optical Absorbance.

Acceptor	E_{m7} *(mV)*	*λ max (nm)*	*Millimolar extinction coefficient* $(10^{-3}\ M^{-1}\ cm^{-1})$	*Molecular weight*	*Soluble in*
DCPIP	+217	590 (oxidised form)	16	290 (disodium salt)	Ethanol
Potassium ferricyanide[a]	+430	420 (oxidised form)	1.04	329	Water
$NADP^+$	−320	340 (reduced form)	6.22	743	Water (most stable at neutral pH)

[a]See ref. 22.

acid to a blue DCPIP solution. Oxidised DCPIP has a high enough extinction coefficient for it to be feasible to measure electron transport as a decrease in red absorbance (at around 590 nm) in a complete reaction mixture with chloroplasts still present. Thus 3 ml of chloroplast (thylakoid) suspension containing 0.1 mM DCPIP in a standard 10 mm cuvette will have an absorbance of about 1.5 at 590 nm relative to a reference cuvette containing no DCPIP. A chloroplast concentration equivalent to 30 μg of chlorophyll will give an absorbance decrease in the region of 0.1 min^{-1} as the DCPIP becomes reduced at reasonable light intensity (e.g. a desk lamp at 50 cm, preferably behind a bottle of water to act as a heat filter). DCPIP, like DCMU, is only sparingly soluble in water, and a stock solution of 0.1 M is best made up in ethanol, taking care to keep the final ethanol concentration in the reaction mixture below 1% (it would be 0.1% in the above example).

Other electron acceptors that can be used in this way are ferricyanide and $NADP^+$. $NADP^+$ is of course the physiological Hill oxidant, and for thylakoid preparations the soluble electron carrier ferredoxin at 5 μM must be added to replace that lost by dilution during the chloroplast isolation. With pea thylakoids it has been found that it is necessary to add ferredoxin-$NADP^+$ reductase as well. A suitable reaction mixture can be devised from the extinction coefficients given in *Table 1*. All three acceptors of *Table 1* are effectively PSI acceptors in intact thylakoids. Ferricyanide and DCPIP will also act as PSII acceptors in sub-chloroplast particles: a kinetic barrier prevents their reduction in intact thylakoids. Mild trypsin digestion (Section 2.5.2) removes this barrier along with the DCMU-sensitivity of their photoreduction. For ferricyanide (absorbing at 420 nm) and NADP (absorbing at 340 nm) it is not usually practical using simple spectrophotometers to measure reduction in the presence of chloroplasts with their strong light-scattering as well as chlorophyll absorbance band at these wavelengths, so it is necessary to spin down the chloroplasts in a bench centrifuge and measure the absorbance change of the supernatant. Ferricyanide and $NADP^+$ reduction can however be measured directly using dual-wavelength spectrophotometry (see Chapter 4).

With fibre-optic actinic illumination it is now quite simple to make continuous measurement of reduction of many electron acceptors, provided care is taken to use an appropriate combination of optical filters to define the actinic beam and to shield the detecting system of the spectrophotometer. Thus for DCPIP reduction a blue actinic beam can be used in conjunction with a red filter covering the entrance to the photomultiplier housing. The absorption band of DCPIP is also sufficiently broad for a long-pass red filter (e.g. a 663 nm cut-off filter which has 50% transmission at 663 nm and transmits more at longer wavelengths; Appendix II) to be used for the actinic beam with the photomultiplier protected by a broad-band blue filter such as a Corning 4-96.

The tendency of chloroplasts and thylakoids to sediment from the suspension during measurement can be overcome by means of an electronic stirring head in the base of the cuvette holder with a magnetic follower ('flea') in the cuvette. An electronic stirrer which fits under a standard 1 cm cuvette is manufactured by Rank Brothers, Bottisham, Cambridge. Use of an electronic stirrer and magnetic 'flea' inevitably generates noise in the absorbance signal, but if the flea is kept below the optical light-path this disadvantage is more than offset by the elimination of drift caused by sedimentation and by the opportunity for thorough mixing of reagents added during the course of the reaction.

2.3 **The Liquid-phase Oxygen Electrode**

Various types of oxygen electrode have been used in photosynthesis research for over 40 years. A rapidly-responding oxygen electrode used with a modulated light source is most useful for measuring rapid transients (8). For steady-state measurements, however, a combined oxygen electrode and reaction vessel is now widely used in photosynthesis research. Commercially produced models are available from Hansatech, King's Lynn, and Rank Brothers, Bottisham, and 'when all else fails, read the manufacturer's instructions' is as sound a principle here as elsewhere. The reader is also referred to the article of Delieu and Walker (9) on whose design the Hansatech electrode is based. The Rank electrode is derived from a design of Chappell, but the same general principles apply since both models are variants of the Clark oxygen electrode, using a platinum cathode and a silver anode connected by a KCl bridge.

2.3.1 *Principles and Construction*

A block diagram of a basic oxygen electrode set-up is given in *Figure 2*. The oxygen electrode chamber is kept at constant temperature by circulating water from a thermostatically-controlled water bath through a transparent water jacket which surrounds the chamber. Electron transport in isolated thylakoids is generally most stable at 15°C, though intact chloroplasts work better at 20 or 25°C. A slide projector is a good, cheap light source but a bottle of water or other heat filter should be placed between the projector and the electrode unit in order to minimise heating effects. Heating should be avoided for two reasons: firstly oxygen solubility is temperature-dependent, and secondly there is the obvious need to keep the reaction mixture at constant temperature. The oxygen electrode disc contains the anode and cathode which are maintained at a constant potential difference by a control box (usually battery powered). The control box also has an output to a chart recorder. A simple potentiometric strip-chart recorder (Y/t recorder) will then record oxygen concentration as a function of time.

Figure 3 shows assembled Hansatech and Rank oxygen electrodes; *Figure 4* shows the electrode discs. In both cases the electrical current passing between the anode and cathode varies linearly with the oxygen concentration in the vicinity of the cathode, where oxygen is reduced electrochemically.

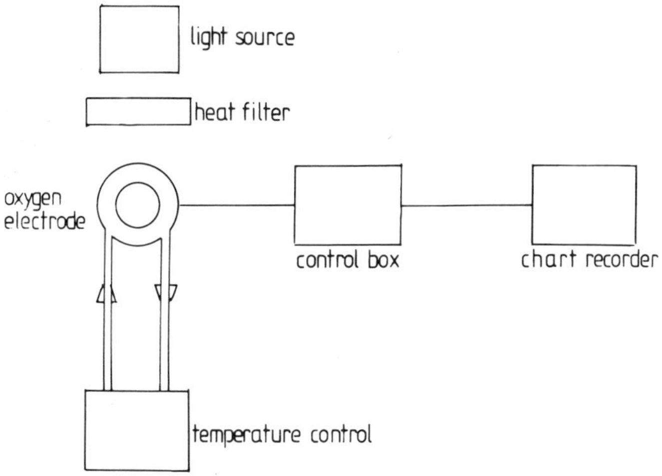

Figure 2. General layout of an oxygen electrode system.

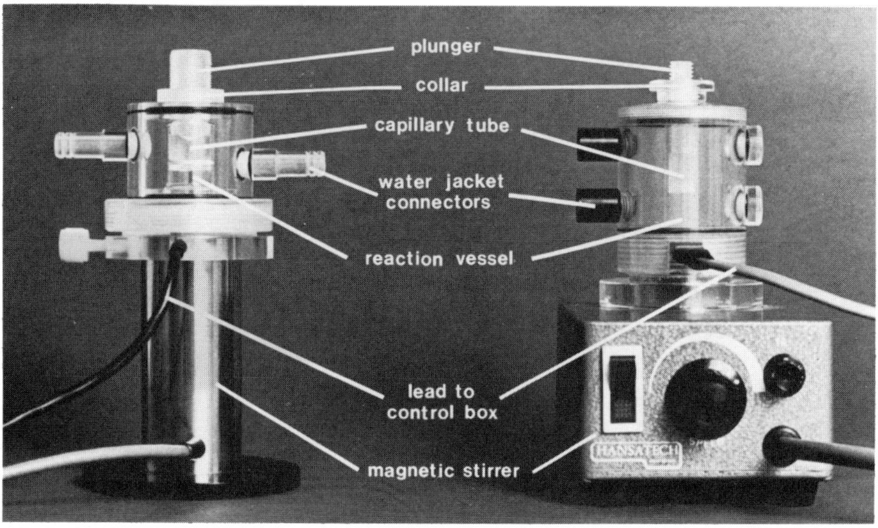

Figure 3. Rank **(left)** and Hansatech **(right)** oxygen electrodes, assembled and mounted on magnetic stirrers.

At the cathode:

$$O_2 + 2H_2O + 2e^- \rightarrow H_2O_2 + 2OH^-$$
$$H_2O_2 + 2e^- \rightarrow 2OH^-$$

At the anode:

$$4\ Ag \rightarrow 4\ Ag^+ + 4e^-$$
$$4\ Ag^+ + [4\ Cl^-] \rightarrow 4\ AgCl$$

The KCl solution in the vicinity of the platinum cathode is separated from the reaction mixture by a thin membrane (usually of Teflon) which is freely permeable to oxygen.

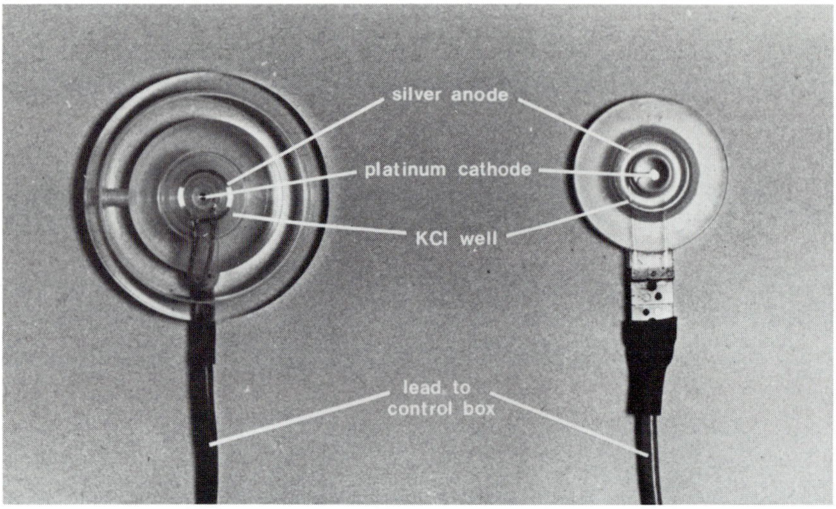

Figure 4. Rank **(left)** and Hansatech **(right)** oxygen electrode discs.

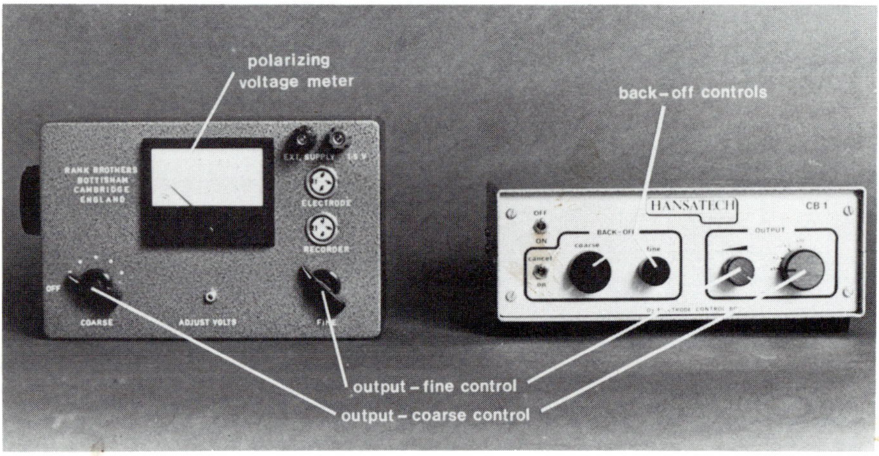

Figure 5. Rank **(left)** and Hansatech **(right)** oxygen electrode control boxes.

The current flowing is therefore proportional to the oxygen concentration in the reaction mixture, with the response time of the apparatus (a few seconds) being limited by diffusion of oxygen from the stirred reaction vessel to the cathode.

The response time decreases with increased cathode surface area [the rapidly responding 'Joliot' electrode used for flash-yield measurements (8) has a large surface area in direct contact with the sample suspension] but increased surface area also increases consumption of oxygen by the cathode itself. For continuous measurements a small cathode is preferred so that electrochemical consumption of oxygen is negligible compared with photosynthetic rate.

Control boxes (*Figure 5*) normally have three basic controls:

(i) A control for the polarising voltage applied between the anode and cathode. This should be set for 650 – 700 mV. The polarising voltage is pre-set on some control boxes.

(ii) A gain or 'output' control (or combination of coarse and fine controls) for adjusting the scale of the output to the chart recorder.

(iii) A zero-suppression or 'back-off' control for adjusting the zero position on the chart recorder without affecting the scale.

It is important that the control box is compatible with the chart recorder. Most control boxes require a fairly sensitive recorder, ideally with a minimum input of 1 mV, whereas one recent control box (Hansatech CB1) will actually give an output of about 1 V. Multiple recorder inputs selected by a switch (say 1, 5, 10, 20, 50, 100 mV, etc.) are useful for changing the expansion factor by a known amount: for example, switching from 10 mV to 1 mV input to give a 10-fold scale expansion.

2.3.2 *Setting Up*

The following instructions for setting up the oxygen electrode are based on those written for well-tried class practical schedules.

(i) Take the electrode disc and cover the silver cathode completely with a few drops of saturated KCl solution. The well in which the anode sits should contain a continuous layer of KCl solution, but should not be over-filled to the extent that KCl solution spills over the rest of the disc.

(ii) Place a further drop of KCl solution on the platinum cathode. Place a small ($\sim$2 cm) square of cigarette paper or lens tissue on the cathode, and a similar sized square of Teflon membrane on top of that. The membrane should not protrude over the edge of the well. The tissue acts as a 'spacer' to maintain a constant layer of KCl solution between cathode and membrane.

(iii) Push a rubber O-ring over the raised projection (containing the cathode) in order to fasten the membrane in position. Avoid trapping air bubbles under the membrane.

(iv) Fasten the disc, now complete with membrane, into the base of the reaction vessel by screwing up the collar. Take care to avoid rotating the disc relative to the reaction vessel as this can tear the membrane. The collar needs to be quite firmly hand-tight so that the rubber ring forms an effective gasket between the reaction vessel and the well in the electrode disc (undertightening is a common fault for beginners).

The Rank oxygen electrode has a single rubber ring which serves as a gasket as well as to hold the membrane in place. The Hansatech oxygen electrode has separate rings for each of these functions, as well as a third ring of larger diameter to prevent leakage of the KCl solution onto the outside of the electrode unit.

(v) Pipette a few millilitres of distilled water into the reaction vessel — if this is left empty for long periods the KCl solution will dry up and the membrane will have to be replaced.

(vi) Connect the electrode to the control box, and the control box to the chart recorder. Place the electrode over a magnetic stirrer and put a magnetic follower or 'flea' in the reaction vessel. Do *not* place the cap or plunger in position yet, the water in the reaction vessel should remain open to the air for equilibration.

(vii) Switch on the magnetic stirrer and set it to give smooth rotation of the flea. Switch on the chart recorder and the water supply to the water jacket. Adjust the zero position on the chart recorder so that the pen records zero deflection with the input selector on zero. Then select an appropriate input sensitivity (usually 10 mV; 1 V on the Hansatech CB1).

(viii) Switch on the control box, and adjust the polarising voltage, if necessary, so that the meter (if present) reads 650–700 mV. Turn the coarse output control clockwise (a few steps if stepped) until the pen on the chart recorder reaches 60–80% of full-scale deflection. The pen will drift slightly and the rate of drift will decrease with time. It should have stopped completely after 10–15 min at the most.

(ix) When the drift has stopped, use a combination of the coarse and fine output controls to bring the pen to read 100%, that is, full-scale deflection. The distilled water in the reaction vessel will now have a known concentration of dissolved oxygen, since it will have equilibrated with the air. The reading of 100% on the chart recorder now corresponds to the concentration of oxygen in air-saturated distilled water at the defined temperature and at normal atmospheric pressure. *Table 2* gives oxygen concentrations at various temperatures. In order for the calibration of the oxygen electrode to remain constant (at 100% deflection for this oxygen concentration) the output controls should not be changed at all during the experiment.

(x) Check the response of the oxygen electrode as follows. Set the chart moving at about 30 mm min^{-1}, drop the pen, and add a few milligrams of solid sodium dithionite (just enough to cover the end of a microspatula) to the water in the reaction vessel. This quickly reduces all the dissolved oxygen. An alternative to adding dithionite is to bubble the water with nitrogen gas from a cylinder. In either case the pen should make an immediate excursion, ideally reaching the zero position on the chart after a minute or so. In practice a reading of less than 10% is acceptable. If the pen does not get this far, add a little more dithionite. If it still does not comply, consult the fault-finding section (Section 2.3.4). Alter-

Table 2. The Oxygen Concentration of Air-saturated Water [from Delieu and Walker (9)].

Temperature (°C)	*O_2 concentration (mM)*
0	0.442
5	0.386
10	0.341
15	0.305
20	0.276
25	0.253
30	0.230
35	0.219

natively, a quick and convenient way of checking that the oxygen electrode is responding without changing the contents of the reaction vessel is to switch off the magnetic stirrer. Consumption of oxygen by the cathode then produces a slow but steady pen excursion which is completely reversed when the stirrer is switched on again.

(xi) Stop the chart and lift the pen. Thoroughly rinse out the dithionite solution and replace it with water. The pen should return to 100%. A vacuum line is the simplest way of removing the liquid contents of the reaction vessel. The end of the line should be a piece of tubing soft enough not to damage the membrane and of internal diameter smaller than the diameter of the magnetic flea.

(xii) If oxygen *evolution* is to be measured in the experiment, the pen will have to be backed-off using the back-off or zero-suppression control on the control box or the zero control on the chart recorder. It is usually advisable to back-off to no less than 10%, to allow for unexpected oxygen uptake. Backing-off may be carried out as necessary during the course of an experiment since it merely alters the position of the oxygen scale on the chart without altering the scale itself. Provided the output controls are not moved, the calibration will be unaffected.

(xiii) Decide on the experimental system, load the reaction vessel with sample and reaction mixture, then secure the cap or plunger in the reaction vessel (the screw collar flush with the top of the water jacket) in such a way that the surface of the reaction mixture rises a few millimetres into the cap's capillary tube; there should be no 'dead' volume (i.e. air bubbles) in the reaction vessel. Additions during the course of a single measurement can be made through the capillary tube by means of a microlitre syringe. For a microlitre syringe with a 50 mm needle the barrel of the syringe can normally be pushed against the top of the plunger without fear of the tip of the needle puncturing the membrane. This procedure ensures that the solution from the syringe enters the bulk of the reaction mixture and quickly becomes stirred by the rotating flea. A common mistake for beginners is to make additions only into the capillary tube of the plunger, presumably from fear of puncturing the membrane.

(xiv) Leave the control box, chart recorder, stirrer and circulating water supply on at all stages during an experiment, even over coffee or lunch-breaks. If it is necessary to leave the electrode for a while between runs, it is a good idea to leave it with water in the reaction vessel, to make sure that the initial baseline has been maintained. Only the chart drive need be switched off, and the pen raised. At the end of a sequence of experiments, the fastidious will appreciate the value of leaving water in the reaction vessel, giving a final opportunity to check the baseline and back-off position. It also helps subsequent setting-up if all controls (including the chart recorder input) are returned to the zero position and the control box switched off.

2.3.3 *Calibration*

Accurate absolute calibration can be carried out by measurement of the release of oxygen from hydrogen peroxide by catalase, though great care has to be taken to standardise the hydrogen peroxide. This method is described by Delieu and Walker (9). In practice the values given in *Table 2* have been found to be highly reliable, and the concentration of dissolved oxygen in air-saturated reaction mixtures is rarely significantly different from those given in *Table 2* for water. However, it is as well to check the calibration of the oxygen electrode at least at the initial stages of an experimental programme, and obviously where absolute values of oxygen yield are crucial to the interpretation of results.

A useful method of calibration under the conditions prevailing in an experiment with isolated thylakoids is to add limiting amounts of freshly made up and standardised potassium ferricyanide solution. A freshly made up solution of 20 mM, when diluted 20 times (e.g. add 150 μl to water in a cuvette to make up to 3.0 ml) will give an absorbance at 420 nm of 1.04 (*Table 1*). 10 μl of 20 mM stock solution (i.e. 200 nmol) added to a suspension of thylakoids in the oxygen electrode will give an increment of light-dependent oxygen evolution equivalent to 50 nmol O_2, since $O_2 = 4\ Fe(CN)_6^{3-}$. This will, for example, correspond to 16.4% of full-scale deflection for a 1 ml reaction mixture at 15°C or 18.1% at 20°C if the oxygen concentrations in *Table 2* are correct. A possible source of error here would be the presence of endogenous or added reductant (such as ascorbate), and it is therefore best to make four or five consecutive additions of ferricyanide, noting the total increment of oxygen evolution in each case. After endogenous reductant has been titrated out, equal volumes of ferricyanide should give equal increments of oxygen if the response of the oxygen electrode is linear. If the standard pen deflection corresponding to each oxygen increment on addition of 10 μl of ferricyanide is y% then the total amount of oxygen in the solution at 100% pen deflection will be

$$\frac{4.81 \times A_{420}}{y} \ \mu\text{mol } O_2$$

where A_{420} is the measured absorbance of the 20-fold diluted stock ferricyanide solution.

2.3.4 *Maintenance and Fault Finding*

Provided electrical connections are all sound and the membrane correctly installed on the electrode disc, oxygen electrodes are surprisingly reliable and even student-proof. Routine maintenance consists simply of cleaning the electrode disc and the base of the reaction chamber free of KCl crystals every time the membrane is changed. Periodically the AgCl deposit, which visibly tarnishes the anode, should be cleaned off with a mildly abrasive metal polish or wadding. The polish should then itself be washed off with detergent. A build-up of AgCl will adversely affect response time. The electrode disc itself should be stored clean and dry (and in the dark — fluorescent striplights discolour the plastic) since KCl deposits can in time cause electrical tracking or short-circuiting. This will lead to a residual signal even in the absence of oxygen. The membrane should be replaced and the disc cleaned of KCl at least once a week if the electrode is in daily use. Replacing the membrane requires a little skill and, particularly for those with long fingernails, a membrane applicator may be a worthwhile investment. The symptom of membrane damage is periodic spikes in the trace which become more frequent and greater in amplitude with time.

Certain reagents such as 2,5-dibromo-3-methyl-6-isopropyl-*p*-benzoquinone (DBMIB) are apparently adsorbed onto the membrane with the result that their effects carry over to subsequent traces. This carry-over effect can be prevented by rinsing the reaction vessel (but not the water jacket!) with ethanol. Periodic oscillations with a frequency of around 1 min may be symptomatic of discharged batteries, but are more likely to reflect temperature changes due to a coarse thermostat setting in the supply to the water jacket. Chattering or oscillations of frequency less than 1 sec may be stirrer noise — this can be checked by varying the stirrer speed. The axis of rotation of an undersized flea may itself rotate, giving much slower oscillations of the trace. In each case the remedy is to replace the flea or to re-position the electrode over the stirrer. Another common stirrer fault is for the speed of rotation of the magnet to increase as the stirrer warms up until the original stirrer setting becomes too fast for the flea to keep up with the magnet.

Any other malfunction of the electrode or recorder can be isolated fairly quickly by checking the response of the electrode [Section 2.3.2 (x), (xi), — just switching the stirrer off will usually do] before and after replacing a suspect component or connector. The success of systematic replacement of components (leads, control box, recorder, electrode disc) depends on there being a very high probability that the replacement is itself in working order. The situation where many components of different kinds have a history of inadvertent damage but have not been repaired or replaced is to be avoided at all costs. Class practicals using such equipment can quickly degenerate into chaos, and frantic replacement of components by others with which the existing electrical connectors are incompatible adds another interesting dimension. In the research laboratory it is probably best to buy your own apparatus and keep it entirely separate from teaching set-ups, for much the same reason than an ex-driving school car is unlikely to provide trouble-free transport.

2.3.5 *Light Sources*

A standard slide projector is the cheapest satisfactory light source. Fibre-optic light sources have some advantages: the lamp may be remote from the oxygen electrode, and the light pipe itself acts as a heat filter. A stabilised power supply is a good idea if a constant light output is essential, though the response of the oxygen electrode itself is slow enough to damp out mains ripple completely.

The wavelength of light used should be defined according to the needs of the experiment, though it is generally sensible to use a broad red filter even for routine work, since low wavelength components of white light can drive purely photochemical oxygen uptake at significant rates, particularly where zwitterionic buffers (e.g. Hepes, MES) are used in the presence of reducing agents and flavins. Ferricyanide and phenazine methosulphate (PMS, the co-factor of cyclic electron flow) are also photosensitive and should be used only with red light unless blue light is required for some other reason (such as excitation of fluorescence).

Light-emitting diodes (LED) of sufficient intensity for photosynthesis have recently become available, and offer the advantages of a stable and well-defined output together with a fast rise time obviating the need for a mechanical shutter.

2.3.6 *Units*

The commonest units for expression of electron transport rates are μmol O_2 (mg chl)$^{-1}$

h^{-1}, typical uncoupled rates at saturating light intensity being several hundred μmol O_2 (mg chl)$^{-1}$ h^{-1}.

It is usually convenient to measure rates initially as pen movement (% of full-scale deflection) per minute, the value being calculated from a linear region of the trace or from a construction line drawn through it. 'Raw' data can be tabulated as '% min^{-1}' and a single conversion factor used for conversion to μmol O_2 (mg chl)$^{-1}$ h^{-1}. This factor is

$$0.6 \times \frac{c.v}{chl}$$

where c is O_2 concentration (mM) at air saturation at that temperature, v is volume of reaction mixture (ml) and chl is the amount of chlorophyll present (mg). For 50 μg of chlorophyll in 1 ml at 20°C the factor is 3.312.

2.4 **Whole-chain Electron Transport**

2.4.1 *Reaction Medium*

A suitable thylakoid reaction medium for most electron transport measurements (including electron transport partial reactions) contains sorbitol (0.1 M), $MgCl_2$ (5 mM), NaCl (5 mM) and Hepes buffer (50 mM) to maintain pH 7.6 (*Table 3*). A stock solution containing twice these stated concentrations can be added to give exactly half the required final volume of reaction mixture; the remainder of the final volume is then made up of the following: water, chloroplast suspension equivalent to the desired total chlorophyll (final concentration usually 50 μg ml^{-1}) and any other reagents that must be added for the particular experiment.

Table 3. Media Recipes.

Thylakoid reaction medium	
	Final concentration
Sorbitol	0.1 M
$MgCl_2$	5 mM
NaCl	5 mM
Hepes (free acid)	50 mM
Adjust to pH 7.6 with dilute KOH	
Medium for oxygen uptake; PSI and PSII (Mehler reaction)	
Sorbitol	0.1 M
$MgCl_2$	5 mM
NaCl	5 mM
Hepes (free acid)	50 mM
Methyl viologen	0.1 mM
Sodium azide	5 mM
Adjust to pH 7.6 with dilute KOH	
Medium for oxygen uptake; PSI	
Sorbitol	0.1 M
$MgCl_2$	5 mM
NaCl	5 mM
Hepes (free acid)	50 mM
DCMU	10 μM
DCPIP	0.1 mM
Ascorbate	5 mM
Adjust to pH 7.6 with dilute KOH	
Intact chloroplast medium	
Sorbitol	0.33 M
EDTA	2 mM
$MgCl_2$	1 mM
$MnCl_2$	1 mM
Ascorbate	2 mM
Hepes (free acid)	50 mM
Adjust to pH 7.6 with dilute KOH	

2.4.2 *Oxygen Evolution*

The physiological electron transport chain can usually be reconstituted by adding ferredoxin (5 μM) and $NADP^+$ (2 mM) to the reaction medium. This system gives net oxygen evolution in the light, but a variable proportion of electron flow to oxygen also occurs, making the system unsuitable for absolute measurement of electron transport unless some correction factor for oxygen reduction is introduced. Oxygen evolution is linear until all the $NADP^+$ becomes reduced, whereupon the oxygen uptake reaction can be observed.

Potassium ferricyanide (5 mM) is a useful electron acceptor that gives light-dependent oxygen evolution dependent on both PSI and PSII acting in series. There is no evidence for competing oxygen uptake. Prolonged incubation of thylakoids with ferricyanide in the dark can impair electron transport and photophosphorylation, and it is therefore preferable to add ferricyanide in a small volume with a microlitre syringe soon after the light has been switched on.

2.4.3 *Oxygen Uptake*

Another method of measuring whole-chain electron transport utilises the autoxidation reaction of a terminal electron acceptor. Using methyl viologen (MV^{2+}) as an example, reduction of methyl viologen by the chain

$$H_2O + 2\ MV^{2+} \rightarrow 2\ MV^+ + ½O_2 + 2H^+$$

gives oxygen *evolution* in the usual stoichiometry of the Hill reaction ($O_2/2e^- = ½$), but oxygen is simultaneously consumed at twice this stoichiometry by a combination of the autoxidation reaction

$$2\ MV^+ + 2O_2 \rightarrow 2\ MV^{2+} + 2O_2^-$$

with dismutation of superoxide

$$2O_2^- + 2H^+ \rightarrow H_2O_2 + O_2$$

The sum of these three equations (electron transport, autoxidation, dismutation of superoxide) is a special case of the Hill reaction, termed the Mehler reaction

$$H_2O + ½O_2 \xrightarrow{(MV)} H_2O_2$$

which gives a stoichiometry of oxygen *uptake* of

$$O_2/2e^- = ½$$

Oxygen is therefore consumed at exactly the same overall rate as that at which it would be evolved in a straightforward Hill reaction. Possible departures from $O_2/2e^- = ½$ are shown in *Table 4*. In order to ensure that this stoichiometry holds, it is necessary to avoid the presence of reducing agents such as ascorbate or Mn^{2+} ions, which will interfere with the dismutation of superoxide, or else to add the enzyme superoxide dismutase to at least 10^3 units ml^{-1}. It is also necessary to inhibit the endogenous catalase activity of the chloroplast preparation (e.g. with sodium azide) to suppress further release of oxygen from H_2O_2, or else to add excess catalase (4×10^3 units) with ethanol (5 mM) whereupon H_2O_2 is consumed by peroxidation of ethanol to

Table 4. Stoichiometries of Oxygen Uptake when Methyl Viologen (MV) is used as an Electron Acceptor.

Enzyme additions		*Reactions involved*	$O_2/2e^-$
Water as electron donor; PSI and PSII			
Without ascorbate	none	(ii)	½
	SOD		
	catalase	(ii) + (iii)	0
	SOD + catalase		
With ascorbate	none	(i)	1½
	SOD	(ii)	½
	catalase	(i) + (iii)	½
	SOD + catalase	(ii) + (iii)	0
Ascorbate as electron donor; PSI only			
	none	(i)	2
	SOD	(ii)	1
	catalase	(i) + (iii)	1
	SOD + catalase	(ii) + (iii)	½

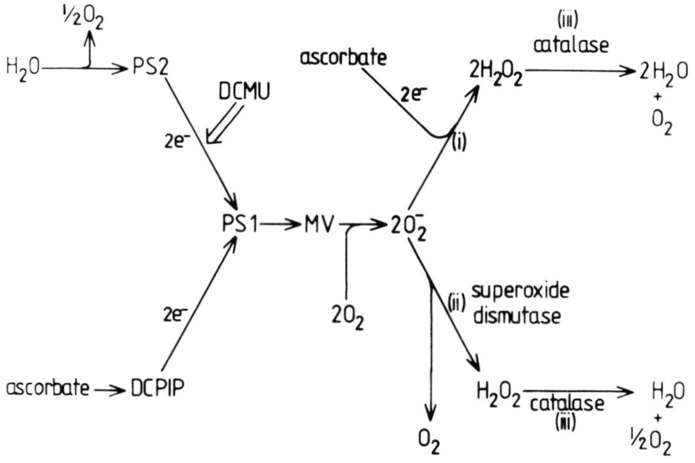

acetaldehyde.

The simplest assay medium for measurement of whole-chain electron transport by the Mehler reaction therefore contains methyl viologen (0.1 mM) and sodium azide (5 mM) in addition to the components given in Section 2.4.1 (see *Table 3*). In a standard oxygen electrode, consumption of oxygen will proceed at a constant rate until 15–20% of the initial oxygen concentration remains, whereupon the apparent rate declines. A number of bipyridylium compounds will mediate photosynthetic oxygen uptake in this way, all at concentrations of 0.1 mM. These include methyl viologen itself ('paraquat'), benzyl viologen, diquat and triquat. The reaction can also be mediated by anthraquinone (0.1 mM), flavin mononucleotide (0.1 mM), adrenochrome (0.1 mM) and by ferredoxin (25 μM) (*Table 5*).

2.4.4 *Photosynthetic Control*

Electron transport is coupled to ATP synthesis (Chapter 6) and it is often useful to be able to assess the degree of coupling in a thylakoid preparation by observing the effect of ATP synthesis on the rate of electron transport. Photosynthetic control can be demonstrated using any whole-chain electron transport system, and some electron transport partial reactions.

The simplest technique is to carry out an initial electron transport measurement (a 'state 2' rate) with K_2HPO_4 (5 mM) also present in the reaction medium and then to add (e.g. after 1 min of linear electron transport) a limiting amount (e.g. 0.2 μmol) of ADP. The effect of ADP on the oxygen electrode trace of well-coupled thylakoids is shown in *Figure 6*. The rate of electron transport increases (to the 'state 3' rate) as thylakoid $\Delta\bar{\mu}_{H^+}$ is dissipated by ATP synthesis; it decreases again (to the 'state 4' rate) once all the ADP has been used up. State 4 is normally slower than state 2 since the ATP synthesised during the cycle serves to block the main channel of proton leakage through the coupling ATPase. The total extra increment of oxygen evolved or consumed, given by B in *Figure 6*, is calculated in μatom or natom equivalents. On the assumption that state 3 is completely a phosphorylating electron transport rate, the number of ATP molecules synthesised per pair of electrons transferred through the chain (the $P/2e^-$ ratio) will be equal to the measured ADP/O ratio, provided $[O] = 2e^-$ (that is, $O_2/2e^- = ½$).

Routine measurement of the ADP/O ratio of thylakoids by these means usually gives a value of 1.33 or thereabouts, though the 'true' $P/2e^-$ ratio is still in dispute. Thus addition of 0.2 μmol (e.g. 10 μl of 20 mM) ADP (A) will normally give an oxygen increment (B) of 150 natom equivalents (75 nmol) of oxygen, corresponding to 24.6% full-scale deflection for a 1 ml reaction volume at 15°C. Where oxygen uptake is measured, it is important not to confuse the transition from state 3 to state 4 with the decline in response of the oxygen electrode at low oxygen concentrations. Addition of NH_4Cl (to 5 mM) after a steady state 4 has been reached should demonstrate a high, linear, uncoupled rate.

A criterion of the integrity of the thylakoid preparation is the photosynthetic control ratio, that is, the ratio of the state 3 rate to the state 4 rate. This is a function of the capacity of the membrane to support a $\Delta\bar{\mu}_{H^+}$ that will exert feedback control on electron transport. The photosynthetic control ratio has a value of three or more in well-coupled thylakoids.

2.5 **Electron Transport Partial Reactions**

2.5.1 *Photosystem I*

For PSI operating alone it is necessary to measure oxygen uptake or reduction of the terminal acceptor since oxygen evolution does not occur. Obviously it is also necessary to add an artificial electron donor in place of water. It is desirable to eliminate any PSII activity that may complicate the results. Thus a simple assay medium for PSI electron transport contains (in addition to components given in Section 2.4.1) DCMU (10 μM) as PSII inhibitor and DCPIP (0.1 mM) with ascorbate (5 mM) as an electron

Table 5. Some Electron Donors and Acceptors, Inhibitors, Mediators etc. for Electron Transport and Redox Measurements.

Reagent	*Function*	*Molecular weight*	*Soluble in*	*Suitable stock concentration*	*Effective final concentration*
ADP (adenosine 5′-diphosphate)		427.2[a]	Water	20 mM	0.2 mM
Anthraquinone	Mediator		Ethanol	Saturated solution	10 μl saturated solution
Ascorbate (sodium salt)	Reductant - donor	198 (anhydrous)	Water	0.5 M	5 mM
ATP (adenosine 5′-triphosphate)		507.2[a]	Water	200 mM	0.2 mM
Benzyl viologen (1,1-dibenzyl-4,4′-bipyridylium dichloride)	Autoxidising acceptor or mediator	409.4	Water	10 mM	0.1 mM
Catalase	H_2O_2 dismutation	—	Water	Crystall. susp.	2500 units ml^{-1}
Cyanide (sodium salt)	Catalase and SOD inhibitor	65	Water	0.5 M (small volume)	5 mM
DBMIB (2,5-dibromo-3-methyl-6-isopropyl-*p*-benzoquinone)	Inhibitor	164.2	Ethanol	1 mM	1–10 μM
DCMU [Diuron; 3-(3,4-dichlorophenyl)-1,1-dimethylurea]	Inhibitor	233.1	>10% Ethanol	10 mM in ethanol diluted to 1 mM in 10% ethanol	20 μM
DCPIP (2,6-dichlorophenol indophenol)	Electron donor (PSI) or acceptor (PSII)	290.1 (Na salt) 268.1 (free acid)	Ethanol	10 mM	0.1 mM
Dithionite	Reductant-donor	210.2 (Na salt) 174.1 (anhydrous)	Water	1 M (in 1 M Tris pH 9.0)	10 mM
DPC (diphenyl carbazide)	Electron donor to PSII	242.3	Ethanol	50 mM	0.5 mM
Duroquinone (tetramethyl-*p*-benzoquinone) ('TMQ')	Electron donor as quinol	164.2	Ethanol	50 mM	0.5 mM
EDTA (ethylene diamine tetra-acetic acid)	Chelating agent	292.25 (free acid)	Water		
FCCP (carbonyl cyanide *p*-trifluoromethoxyphenyl-hydrazone)	Uncoupler	254	Ethanol	1 mM	10 μM
Ferredoxin	Electron carrier; autoxidisable acceptor	—	—	—	5–20 μM
Ferricyanide (K salt)	Electron acceptor; oxidising agent	329.3	Water	0.2 M	2 mM
Ferrocyanide	Donor	329.3	Water	50 mM	0.5 mM
FMN (flavin mononucleotide)	Autoxidising acceptor	456	Water	10 mM	50 μM

Table 5. continued.

Reagent	*Function*	*Molecular weight*	*Soluble in*	*Suitable stock concentration*	*Effective final concentration*
Hepes (N-2-hydroxyethylpiperazine-*N'*-2-ethane sulphonic acid)	Buffer	238.3 (free acid)	Water		
Hydroxylamine	Donor	33	Water	50 mM	0.5 mM
Methyl viologen (1,1'-dimethyl-4,4'-bipyridylium dichloride)	Autoxidising electron acceptor; mediator	257.2[a]	Water	5 mM	50 μM
MOPS [3-(N-morpholino) propanesulphonic acid)]	Buffer	209.3 (anhydrous)	Water		
NADP (nicotinamide adenine dinucleotide phosphate)	Acceptor	765.4[a]	Water	100 mM	1 mM
$NADPH_2$ (reduced form)		833.4[a]			
Oxaloacetate	Acceptor		Water	100 mM	1 mM
PD (*p*-phenylene diamine)	Acceptor/donor	108.1	Ethanol	10 mM	0.1 mM
3-Phosphoglycerate (disodium salt)	Acceptor	230[a]	Water		
PMS (N-methyl phenazonium methosulphate)	Cyclic co-factor	306	Water	10 mM	0.1 mM
Semi-carbazide	PSII donor		Ethanol	50 mM	0.5 mM
SOD (superoxide dismutase)	O_2^- dismutation	—	—	—	—
Sodium azide	Catalase inhibitor	65.02	Water	1 M	5 mM
Sodium borohydride	Reducing agent	37.8	Add solid		
Sodium pyrophosphate	Inhibitor of Pi translocator	446.1 (decahydrate)	Water	500 mM	5 mM
TMPD (tetramethyl-*p*-phenylene diamine)	Mediator	164.3	Water	10 mM	0.1 mM
Tris (Tris hydroxymethyl methylamine)	Buffer	121.14	Water	As required	

[a]These molecular weights are for anhydrous reagents. The amount of H_2O combined varies, refer to suppliers' data sheet for exact molecular weights.

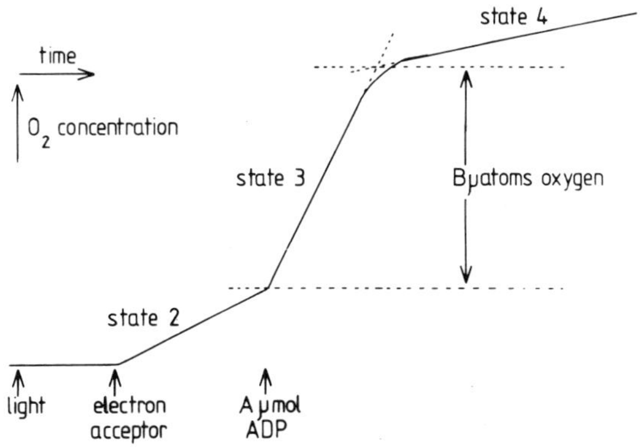

Figure 6. Oxygen electrode trace obtained with isolated thylakoids in an ideal photosynthetic control cycle. ADP/O ratio = A/B. Photosynthetic control ratio = state 3/state 4. K_2HPO_4 (at 5 mM) is present at the start and ADP (A μmol) (e.g. A = 0.2 μmol) is added where shown. NH_4Cl (to give 5 mM) is added during state 4.

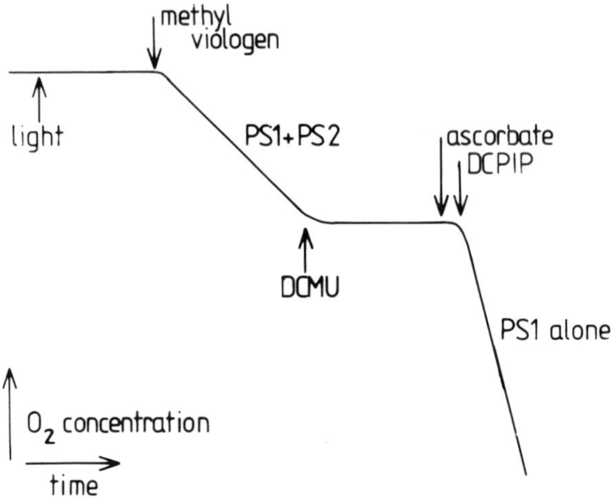

Figure 7. Oxygen electrode trace obtained during measurement of electron transport through PSI and PSII together (following addition of methyl viologen) and of PSI alone (following addition of ascorbate and DCPIP to DCMU-inhibited thylakoids). Final concentrations: methyl viologen, 0.1 mM; DCMU, 10 μM; DCPIP, 0.1 mM; ascorbate, 5 mM. Sodium azide (5 mM) is present throughout.

donor couple (DCPIP effectively mediating electron transport from ascorbate to P700), see *Table 3*.

Electron transport may then be measured as $NADP^+$ reduction at 340 nm (Section 2.2.3) or as oxygen uptake in an oxygen electrode with methyl viologen (0.1 mM) and sodium azide (5 mM) (Section 2.4.3). The oxygen electrode in fact provides a simple way of screening thylakoids for blocks in electron transport (due to chemical inhibition or mutation, for example) associated with PSI or PSII (*Figure 7*). A block in PSII will give a decreased rate of initial oxygen uptake (PSI + PSII) but not of oxygen uptake due to PSI alone. A block in PSI will decrease both rates.

A number of acceptors can be used in place of methyl viologen (Section 2.4.3) and donating mediators in place of DCPIP (Section 2.5.3).

Although the ascorbate → DCPIP → PSI → MV → O_2 pathway provides a satisfactory *relative* measure of PSI electron transport, a number of additional factors must be controlled if it is used as an *absolute* measure. In the simplest system, free of catalase or superoxide dismutase, the $O_2/2e^-$ ratio has a value of 2 (10). This value is obtained because ascorbate reduces the superoxide intermediate of oxygen reduction to peroxide, as well as functioning as an alternative donor to water. Thus the component reactions are:

(i) electron transport (two electrons):

$$AscH_2 + 2MV^{2+} \xrightarrow[PSI]{DCPIP} Asc + 2MV^+ + 2H^+$$

(ii) autoxidation of methyl viologen:

$$2MV^+ + 2O_2 \rightarrow 2MV^{2+} + 2O_2^-$$

(iii) reduction of superoxide to peroxide:

$$2O_2^- + AscH_2 + 2H^+ \rightarrow 2H_2O_2 + Asc$$

The sum of these three reactions is:

$$2\ AscH_2 + 2O_2 \xrightarrow[PSI]{DCPIP,\ MV} 2Asc + 2H_2O_2$$

and hence the $O_2/2e^-$ ratio has a value of 2. In the presence of superoxide dismutase the dismutation reaction:

$$2O_2^- + 2H^+ \rightarrow H_2O_2 + O_2$$

will compete with reduction of superoxide by ascorbate, giving a changed overall stoichiometry of $O_2/2e^- = 1$. Thylakoids can normally be washed free of competing superoxide dismutase activity and, if in doubt, the rate of oxygen uptake can be titrated with added superoxide dismutase: the rate should be exactly halved by saturating superoxide dismutase if the enzyme was absent initially. Alternatively, potassium cyanide (10 mM) can be added to inhibit superoxide dismutase.

Similar considerations apply to contaminating catalase, though this can be inhibited with sodium azide (5 mM). At these concentrations, cyanide (10 mM) has been found to inhibit both catalase and superoxide dismutase, while azide (5 mM) inhibits only catalase (10). These effects on oxygen uptake stoichiometry are illustrated by the trace shown in *Figure 8*. *Table 4* shows $O_2/2e^-$ ratios prevailing for all oxygen-consuming electron transport measurements of the kind outlined here. The simplest protocol is probably to use well-washed thylakoids with 5 mM sodium azide and to assume $O_2/2e^- = ½$ for PSI and PSII together and $O_2/2e^- = 2$ for PSI alone.

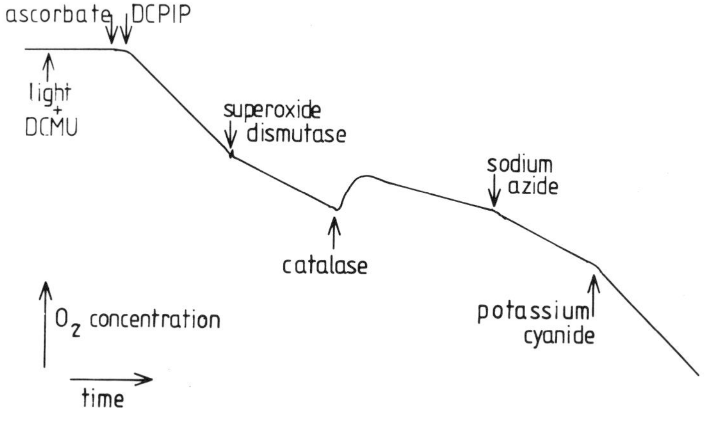

Figure 8. Oxygen electrode trace obtained during measurement of PSI electron transport as in *Figure 7*. Superoxide dismutase (800 units) halves the observed rate of oxygen uptake, and this rate in turn is halved by addition of catalase (4×10^3 units). Sodium azide (to 10 mM) inhibits catalase and potassium cyanide (to 10 mM) inhibits both catalase and superoxide dismutase. The rate of electron transport is constant throughout the experiment; only the stoichiometry of oxygen uptake is changed. See ref. 10 for a more complete description.

2.5.2 *Photosystem II*

In intact thylakoids the lipophilic electron acceptors diaminodurene (DAD) and *p*-phenylenediamine (PD) are useful for measurement of electron flow through PSII alone. They should be freshly made up in buffered medium containing excess ferricyanide at three times the stock concentration of DAD or PD. A suitable final concentration of the acceptor couple is then DAD or PD (0.4 mM) with ferricyanide (1.2 mM). PSI must be inhibited, for example by DBMIB (1 μM), or by pre-incubation for 60 min with KCN (30 mM). Mediators of PSII oxygen uptake have been reported (11), including DBMIB at higher concentrations (20 μM) than those used for electron transport inhibition (1 μM).

In the absence of lipophilic electron acceptors such as DAD or PD, DCPIP and ferricyanide will function as PSII electron acceptors only if the kinetic barrier to their reduction is removed. This is likely to be the case in sub-chloroplast particles. In intact thylakoids, mild trypsin digestion (80 μg ml^{-1} for 10 min at neutral pH in the dark) reveals a DCMU-insensitive electron flow to ferricyanide that may be measured as oxygen evolution. This effect presumably results from removal of the lipophilic shield represented by the Q_B (Chapter 1) or herbicide-binding protein, though the list of likely side-effects of trypsin digestion is long and somewhat limits the usefulness of this technique. Similar reservations apply to the use of silicomolybdate, which also gives DCMU-insensitive PSII electron transport and oxygen evolution. Silicomolybdate can be made up as a stock solution of 25 mM in 50% dimethylsulphoxide (DMSO), and a final concentration of 1 μM will give limited PSII electron transport and coupled photophosphorylation.

2.5.3 *Further Donors, Acceptors and Inhibitors*

More specific localisation of electron transport partial reactions is possible, and comprehensive accounts of these are given in the reviews of Trebst (12,13) and Izawa (14).

(i) *PSI donors other than DCPIP.* One problem that arises from the use of DCPIP/ascorbate as a PSI electron donor (Section 2.5.1) is that very little of the intermediary electron transport chain is involved, as shown by the insensitivity of this reaction to DBMIB, an inhibitor of plastoquinone oxidation. It is often useful when working with thylakoids to use donors to specific points of the chain prior to P700. Duroquinol, for example, functions as a donor to plastoquinone and gives a DBMIB-sensitive PSI electron transfer reaction. Duroquinone is kept as a stock solution (50 mM in ethanol). Prior to use, mix 2 mg of sodium borohydride with 0.5 ml of the stock solution, and leave the mixture on ice until the yellow duroquinone goes colourless on its reduction to duroquinol. Add 5 μl of concentrated HCl to remove excess borohydride. Addition of the duroquinol to a final concentration of 0.5 mM gives DBMIB-sensitive PSI electron transport in intact thylakoids. The difference between the sites of donation by duroquinol and DCPIP is indicated by the oxygen electrode trace in *Figure 9*. The oxygen uptake stoichiometry of the duroquinol−PSI reaction has not been investigated, though it is likely that the same considerations hold as for DCPIP/ascorbate (Section 2.5.1), and hence that $O_2/2e^- = 2$. The electron transport sequence duroquinol → PQ → PSI → MV → O_2 is also coupled to photophosphorylation, presumably because the native

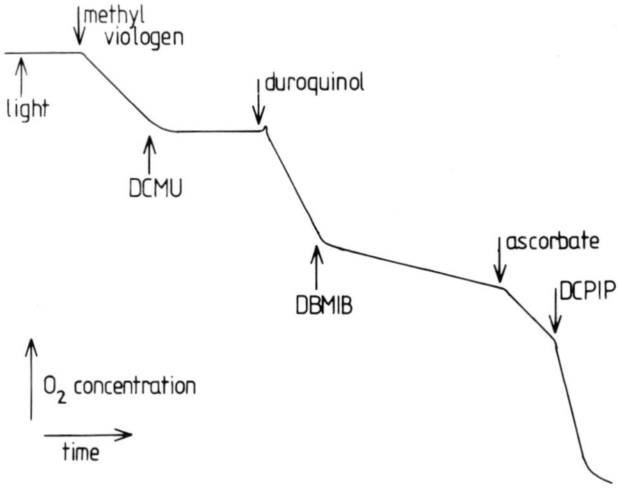

Figure 9. Oxygen electrode trace illustrating further dissection of the non-cyclic chain (cf. *Figure 7*). Duroquinol (to 0.5 mM) donates to plastoquinone and therefore bypasses inhibition by DCMU (5 μM) but not by DBMIB (20 μM). DBMIB is an inhibitor of plastoquinone oxidation. Ascorbate (5 mM) with DCPIP (0.1 mM) bypasses inhibition by DBMIB. From ref. 24.

proton-translocating cycle is retained. DCPIP/ascorbate also gives some coupled photophosphorylation, though photosynthetic control is difficult to observe.

Other electron donors to PSI can be used with ascorbate in place of DCPIP. Tetramethyl-*p*-phenylenediamine (TMPD) gives fast electron transport that is not coupled to ATP synthesis and which is DBMIB-insensitive since it bypasses the plastoquinone proton shuttle. DAD/ascorbate gives coupled electron transport also insensitive to DBMIB — in this case it is likely that DAD catalyses an artificial proton shuttle, replacing plastoquinone.

(ii) *Artificial PSII donors.* In general, PSII donors replace water only where water-oxidation has been inactivated by some means, for example by heating (6 min at 55°C), by Tris treatment (0.8 M Tris, pH 8, for 15 min) or by hydroxylamine treatment (5 mM hydroxylamine at pH 7 for 15 min).

Membranes prepared in one of these ways will then oxidise a variety of artificial donors, including ascorbate, hydroquinone, semicarbazide, ferrocyanide, hydrogen peroxide and PD. The electron transport is DCMU-sensitive. 1,5-Diphenylcarbazide (at 0.5 mM) is probably now the most widely used electron donor to PSII. Since oxygen evolution is absent in these systems, DCPIP reduction measured at 600 nm (Section 2.2.3) is probably the best technique for following the course of the reaction, though PSI acceptors (e.g. $NADP^+$ with ferredoxin) can also be used.

2.6 Measurements with Intact Chloroplasts

2.6.1 *Reaction Conditions*

An intact chloroplast preparation offers a more 'physiological' system for electron transport measurement, but it suffers the drawback of the impermeability of the chloroplast envelope to many added donors, acceptors and inhibitors. Nevertheless, it is often desirable to monitor the complete photosynthetic electron transport chain in this way. Conditions for isolation and storage of intact chloroplasts are given in Chapter 2. A reliable isotonic reaction medium for intact chloroplast studies (*Table 3*) contains sorbitol (0.33 M), EDTA (2 mM), $MgCl_2$ (1 mM), $MnCl_2$ (1 mM), ascorbate (2 mM) and Hepes (50 mM) at pH 7.6. The function of EDTA in the presence of Mg^{2+} and Mn^{2+} is unclear, but this particular medium is nevertheless widely used.

In other respects an intact chloroplast preparation can be used in the liquid-phase oxygen electrode in much the same way as a suspension of thylakoids. It is usually desirable to increase the chlorophyll concentration to 100 μg ml^- and to use a higher temperature of 20 or 25°C. Catalase at about 5×10^3 units is often added to the chloroplast suspension to prevent inhibitory effects of any hydrogen peroxide produced by oxygen reduction.

2.6.2 *CO_2-dependent Oxygen Evolution*

Saturating CO_2 concentrations can be produced by inclusion of sodium bicarbonate (10 mM). In the presence of K_2HPO_4 (0.5 mM), high quality intact spinach chloroplasts will then show physiological rates of oxygen evolution, after an initial lag phase, of 200 μmol O_2 (mg chl)$^{-1}$ h^{-1} or even more. With isolated pea chloroplasts (ideally from 10-day-old pea seedlings) it is necessary to replace the K_2HPO_4 with sodium pyrophosphate (5 mM) and ATP (1 mM), apparently to discourage an adenylate translocator on the chloroplast envelope which will otherwise inhibit photosynthesis by depleting the chloroplast of ATP.

The electron transport measured in such a system is obviously sensitive to DCMU and, less obviously, to uncouplers [e.g. carbonyl cyanide *p*-trifluoromethoxyphenylhydrazone (FCCP) at 1 μM or NH_4Cl at 40 mM]. The action of uncouplers is indirect: inhibition of ATP synthesis inhibits recycling of $NADP^+$ by the coupled phosphoglycerate kinase and triose phosphate dehydrogenase reactions, and effectively removes the terminal electron acceptor. The Calvin cycle is generally thought to require ATP and NADPH in the ratio 3:2, and it is a matter of debate whether the non-cyclic electron transport chain alone provides ATP with a sufficiently high stoichiometry for CO_2 fixation.

2.6.3 *Other Oxygen-evolving Reactions of Intact Chloroplasts*

It is also possible to replace $NaHCO_3$ with 3-phosphoglycerate (1 mM) or with oxaloacetate (1 mM). In both cases the full non-cyclic electron transport chain operates, but these reactions are of use because of their differing stoichiometric requirements for ATP. 3-Phosphoglycerate-dependent oxygen evolution results from regeneration of $NADP^+$ by an isolated section of the Calvin cycle, and requires ATP and NADPH in equimolar amounts. The reaction, like CO_2-dependent oxygen evolution, is sensitive to inhibition by uncouplers, but any auxiliary photophosphorylation (cyclic or pseudocyclic) is probably not required. Oxaloacetate reduction requires no ATP and oxaloacetate-dependent oxygen evolution is therefore stimulated by uncouplers.

2.7 Combining Electron Transport with Other Measurements

2.7.1 *Sampling*

Liquid-phase oxygen electrodes of the type described in Section 2.3 afford the opportunity of withdrawing samples of the reaction mixture at intervals for further analysis, thereby allowing direct comparison of the oxygen trace with a plotted time-course of some accompanying, related reaction. Examples of interesting reactions that can be followed are reduction of a terminal electron acceptor (e.g. $NADP^+$), ATP synthesis (using ^{32}P incorporation or the luciferin-luciferase assay, Chapter 6), protein phosphorylation [each sample being run on sodium dodecylsulphate-polyacrylamide gel electrophoresis (SDS-PAGE) and the gel autoradiographed and bands counted], CO_2 fixation (incorporation of ^{14}C into acid-state products), and concentrations of metabolic intermediates (a suitable assay being carried out on each sample). Samples obviously may also be withdrawn for spectroscopic analysis, e.g. injection of the sample into a tube which is then immersed in liquid nitrogen for fluorescence or e.p.r. spectroscopy (Chapter 4).

Provided there is uniform illumination of the reaction vessel, it does not matter if the total volume of sample taken is an appreciable proportion of the initial volume of reaction mixture — the trace continues to record the rate of change of oxygen concentration, which is independent of volume. An event marker on the chart recorder is particularly useful where sampling techniques are used. Each sample should be taken with a microlitre syringe of the appropriate capacity, and with a needle length (50 mm for Rank and Hansatech oxygen electrodes) that will allow rapid removal of a sample from the centre of the reaction vessel without any danger of puncturing the membrane. For work with radioisotopes, it is convenient to keep one oxygen electrode unit specifically for this purpose.

2.7.2 *Continuous Measurement of Other Variables*

A modification introduced by Horton (15) of the Delieu and Walker oxygen electrode has been incorporated into a unit produced commercially by Hansatech. This consists of an oxygen electrode completely enclosed within an opaque case with five ports for illumination or detection of light. This oxygen electrode permits continuous optical or fluorescence measurement during the time-course of an electron transport reaction. Light can be provided by fibre-optics or by an LED array plugged directly into one of the ports. This type of oxygen electrode has been most extensively used for combined chlorophyll fluorescence measurements. Here photodiode detectors can be plugged directly into one or more ports or a photomultiplier can be mounted close to each port, or at the end of a light-pipe.

This oxygen electrode unit has a number of other potential applications, including measurement of thylakoid ΔpH by 9-aminoacridine fluorescence quenching (Chapter 6) or, indeed, any other fluorescence signal.

It is also possible to measure light scattering and optical absorbance (Chapter 4) using suitable light sources, detectors and filter combinations. Ambitious experiments using four or more channels in addition to the oxygen signal are possible. Oxygen electrode 'polygraphy' of this kind is most easily obtained using a multi-channel chart recorder, and this may be facilitated by the use of recorders (e.g. certain Rikadenki models) which can store one or more channels in order to present the signals onto the chart without the usual offset on the time-axis between channels. Alternatively, a suitably programmed microcomputer offers enormous advantages for the economical storage and subsequent manipulation of the data, though at the time of writing there is no commercially available package designed for this specific application.

2.8 **The Gas-phase Oxygen Electrode**

Photosynthetic electron transport can also be measured as oxygen evolution by leaf discs. A commercially available gas-phase electrode (Hansatech) based on the design of Delieu and Walker (16) works on the same principle as the liquid-phase oxygen electrode but the reaction vessel is replaced by a chamber which allows space (~5 ml) for a leaf disc (up to 25 mm diameter). Capillary matting is also provided to hold a small volume of bicarbonate solution as a source of CO_2. Calibration is achieved by altering the oxygen partial pressure by inserting or removing a known volume (e.g. 1 ml) of air. The most useful version of the gas-phase electrode is likely to be the cased version equipped with ports for simultaneous fluorescence measurement.

The major advantage of leaf discs is that they circumvent all the biochemical problems of obtaining functional particles such as chloroplasts in suspension, and offer an electron transport system functioning essentially the same as *in vivo*. They are likely therefore to be of most use where it is necessary to make simple comparisons of overall photosynthetic rate, perhaps between different species or stages of leaf development. The disadvantages are that the electron transport system that leaf discs contain is almost immune to conventional, invasive, biochemical analysis. Introduction of well-defined photosynthetic inhibitors that can penetrate leaf discs in the gas phase would do something to offset this disadvantage, though tortuous diffusion paths and slow response would still presumably make the liquid-phase electrode and isolated chloroplasts indispensable for serious biochemical analysis of electron transport and energy coupling in plants.

The trade-off between assurance of physiological conditions *in vivo* and simplicity and intelligibility of components isolated *in vitro* is a permanent dilemma in biology. Partial circumvention of this dilemma is the chief advantage of making non-invasive, biophysical measurements on intact tissue as well as on well-defined biochemical model systems. Chlorophyll fluorescence, for example, can report on fundamental events in photosynthesis even in leaves or leaf discs (Chapter 4), and simultaneous measurement of chlorophyll fluorescence yield and electron transport has much to commend it at all levels of complexity.

3. REDOX TITRATION

3.1 **Oxidation-reduction Potentials**

An appreciation of the energetics of the constituent reactions of photosynthetic electron flow is important in understanding the process of photosynthetic energy conservation as a whole. Because the pathway comprises a series of reduction-oxidation reactions (redox reactions) the thermodynamic relationships involved can be conveniently described in electrochemical terms.

The following is only a brief description of the thermodynamics of electron transfer systems; for a fuller account the reader is referred to references 17–19.

A generalised redox reaction:

$$A_{red} + B_{ox} \rightleftharpoons A_{ox} + B_{red}$$

can be considered as the sum of two 'half-cell' reactions of the type

$$A_{red} \rightarrow A_{ox} + ne^-$$

where A_{red} and A_{ox} together constitute a 'redox couple' and n indicates the number of electrons involved.

For such a half-cell, the relationship between the free energy available and the activities of the reactants is given by:

$$E = E_o + \frac{2.3\ RT}{nF} \log_{10} \frac{A_{ox}}{A_{red}} \qquad \text{Equation 1}$$

where E is the redox potential, E_o the standard redox potential, F the Faraday Constant, R the Universal Gas Constant, T the temperature in K, and A_{ox} and A_{red} the activities, or more usually the concentrations, of oxidised and reduced forms of the couple, respectively. This is directly analogous to the familiar Gibbs Free Energy relationship with $E_o = -nFG^o$, where G^o is the standard free energy. At room temperature (25°C) the value of $\frac{2.3\ RT}{F}$ is 59 mV.

Redox potentials have no absolute value and are therefore expressed relative to a standard half-cell, usually the standard hydrogen electrode,

$$H_2 \rightleftharpoons 2H^+ + 2e^-$$

which is given a redox potential $E_o = 0$ mV.

The expressions used above apply to standard conditions at pH 0. It is more useful, with biological systems, to refer to standard conditions at pH 7; a standard redox potential defined at pH 7 is symbolised E_{m7} or sometimes E_o'. Similarly, when referring to standard conditions at pH 7, E becomes E'. The E_o' for the standard hydrogen electrode is −420 mV. The term E_h, frequently encountered in the biochemical literature, signifies that the redox potential is expressed relative to the standard hydrogen electrode.

Therefore the version of Equation 1 usually encountered in bioenergetics is:

$$E_h = E_{m7} + \frac{59}{n} \log_{10} \frac{A_{ox}}{A_{red}} \qquad \text{Equation 2}$$

E_{m7} is frequently referred to as the mid-point potential and is that redox potential where the concentrations of oxidised and reduced forms of the couple are equal. Thus for the generalised redox reaction above, with two participating half-cells

$$\Delta E_h = \Delta E_{m7} + \frac{59}{n} \log_{10} \frac{A_{ox}.B_{red}}{A_{red}.B_{ox}} \qquad \text{Equation 3}$$

where ΔE_h and ΔE_{m7} are the difference between the parameters, E_h and E_{m7}, for the two half-cells; n being the number of electrons involved and A_{ox}, A_{red}, B_{ox}, B_{red} the activities, or in most practical applications, the concentrations, of the reactants. For a more extensive discussion of this relationship see references 17 and 18.

In the situation where either the oxidised or the reduced form is preferentially bound by another chemical species, the apparent mid-point potential will shift depending on the concentration of the ligand present. The analysis of such interactions is discussed in (18). The involvement of protons in bioenergetics gives particular importance to the case where the ligand is a proton bound to the reduced form of the couple, i.e:

$$A_{ox} + e^- + H^+ \rightleftharpoons A_{red} \qquad \text{Equation 4}$$

The mid-point potential then depends on the pH in the following manner:

$$E_{mpH} = E_{m7} + 59\ (7\text{-pH}) \qquad \text{Equation 5}$$

where E_{mpH} is the mid-point potential at the given pH.

When the proton is bound to the oxidised form of the couple, the relationship is similar but the E_{mpH} increases rather than decreases with pH.

3.2 What Redox Titrations Tell You

Measurement of redox potential allows components of electron transport chains which have the same or similar absorbance spectra (or e.p.r. spectra) to be resolved. This is particularly useful when several species of the same component are present in a membrane. Once they have been resolved on the basis of redox potential it is often possible to show small differences in their absorbance spectra (19). In addition, equilibrium measurements of redox potential show the sequences of electron transport components in a membrane that are thermodynamically feasible, under given ambient conditions such as pH. However, for a true picture of the reaction pathways involved, the kinetics of oxidation and reduction of the components can be measured at different redox potentials. As photosynthetic electron flow is initiated by a photochemical reaction, it is technically a fairly simple matter to establish a given ambient redox potential, excite the sample with flash illumination and then observe electrons flowing through (at least) some of the components. In a simple linear sequence of optically detectable components, of course, a series of complementary transient reductions and oxidations would be seen at a potential optimal for electron flow. In fact, real life is more complex than this; nevertheless, considerable progress has been made over the last 15 years in understanding photosynthetic systems using this approach.

At this point, it is worth mentioning that the mid-point potential is not necessarily the working potential of the component. Little information is, as yet, available on the working poises of redox components in photosynthesis, and presumably they will change with varying conditions. The redox poise of any one component will depend on that of other redox components in the chain, and also on other membrane phenomena, such as the protonmotive force.

It has also proved possible, by redox poising, to titrate other reactions in bioenergetic membranes such as proton binding, electric field-indicating absorbance changes and the phosphorylation of light-harvesting pigment-protein complexes, and to correlate these with electron transport events.

3.3 Measuring Oxidised and Reduced Components

To measure the mid-point potential, E_{m7}, of a component it is necessary to be able to measure the ambient redox potential, E_h, and the degree of oxidation (or reduction) of the component of interest simultaneously. Equation 2 can then be used to calculate a value for E_{m7} under those conditions.

For components which have oxidised and reduced forms with significantly different optical extinction coefficients at a convenient wavelength, usually in the visible or near i.r. regions of the spectrum, absorbance spectrophotometry is the method of choice (Chapter 4). A major problem encountered with spectrophotometry in photosynthetic systems is that the relatively massive absorbances of the chlorophyll and carotenoid pigments give a high background absorbance often at the wavelengths where the redox components absorb. In the case of purple non-sulphur bacteria, several mutant strains exist, with shifted carotenoid spectra, which leave the α-bands of the cytochromes visible.

Even when the sample is carefully balanced initially with a reference sample, problems can still arise, particularly if the sample is left in the instrument for some time. It is useful to have a spectrophotometer with the facility for shutting off the measuring beam between measurements. The length of time the sample spends at high or low extremes of potential should be minimised. Chlorophylls can become oxidised at high potentials, bleaching the sample.

Most spectrophotometers of adequate precision can be used for redox titrations, although minor modifications may be necessary to house the cuvette and to allow additions to be made to the solution during the titration. Scanning spectrophotometers have some advantages over double-beam instruments and can be used to titrate more than one component (for example *b*- and *c*-type cytochromes) in a single run. The addition of an on-line computer can eliminate tedious analysis of the results.

For components without convenient optical absorbance signals, the use of other measuring techniques is described in the literature. Electron paramagnetic resonance has been widely used to monitor quinones and iron-sulphur centres, including the components on the acceptor side of both photosystems. It is necessary to freeze the sample rapidly in order to see the e.p.r. signal and an apparatus is described in (19) for doing this and for freezing-in the redox state of the components. Redox potential is sensitive to temperature, but the methods available for freezing appear to monitor faithfully the room temperature equilibrium (see ref. 19 for a discussion of this point).

Other techniques used with redox titrations include chlorophyll fluorescence yield (to indicate the oxidation state of reaction centre components; see Chapter 4) and phosphorylation of light-harvesting pigment-protein complexes (monitoring the activation of a protein kinase).

It is possible to get functional information about the behaviour of redox components by following the kinetics of the changes which occur after a brief flash of light (preferably eliciting only a single turnover). In this type of experiment, the potential is poised as required and then electron transport started by flash excitation. The transient oxidation/reduction of components can be followed spectrophotometrically. The system is equilibrated with the electrode before and after, but clearly not during, the transient. A picture of the behaviour of each visible component can then be followed over the range of potentials at which the chain works, and the properties of components not visible spectrophotometrically can be inferred from the behaviour of the visible components. These invisible components may then be correlated with components seen using other techniques, e.g. e.p.r.

3.4 **Measuring Redox Potential**

Ambient redox potential is followed by an inert metal electrode immersed in the solution; platinum is generally used as the electrode. The potential is measured by comparison with a reference electrode using a high resistance voltmeter. The standard hydrogen electrode, to which potentials are referred, is, in practice, substituted by either a calomel or a silver/silver chloride reference electrode.

The calomel electrode consists of mercury together with a paste of mercury and mercurous chloride (calomel) in contact with a solution of potassium chloride saturated with mercurous chloride, the half-cell reaction is:

$$\tfrac{1}{2}Hg_2Cl_2 + e^- \rightleftharpoons Hg + Cl^-$$

and the mid-point potential is +244 mV at 25°C with saturated KCl.

The silver/silver chloride electrode consists of silver in contact with silver chloride in 1 M potassium chloride. It has a half-cell reaction:

$$AgCl + e^- \rightleftharpoons Ag + Cl^-$$

and a mid-point potential of +222 mV at 25°C.

When calculating values for E_h, the potential of the reference electrode must, of course, be taken into account.

3.5 **Experimental Assembly**

The apparatus for carrying out redox titrations of optically detectable components is described below. It consists of an enclosed cuvette, incorporating an electrode and a stirring device, through which nitrogen or argon is bubbled to maintain anaerobic conditions. Many standard pH meters can be used to record the potential, with digital meters being easy to read.

3.5.1 *Electrodes*

The measuring electrode is made of an inert metal, usually platinum, although gold electrodes can also be used. Commercially, platinum electrodes are available on their own or in combination with a silver/silver chloride reference electrode. For use in a cuvette, where space is a limiting factor, a micro-combination electrode is the most suitable size.

As an alternative to a commercial reference electrode, a calomel electrode outside the cuvette, but connected with the buffer via a salt bridge, can be used. This is less convenient, since the salt bridge has to be made up each time, but it occupies less space in the cuvette and can be used also with a top stirrer. The buffer from the cuvette is brought into electrical contact with the potassium chloride in the reference electrode via a three-way glass tap.

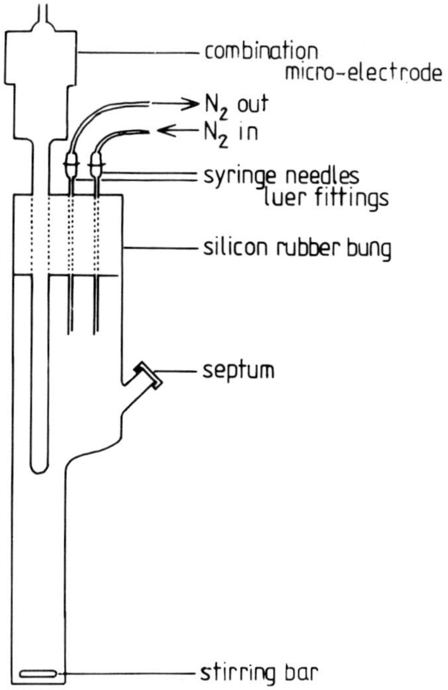

Figure 10. Cuvette for redox titrations (see ref. 19).

3.5.2 *Cuvette*

The cuvette can be made entirely of glass with the electrode and gas inlet/outlet supported in a rubber stopper as in *Figure 10*. The inlet port for making additions at the side should be covered with a septum, sufficiently soft to allow the repeated entry of a microsyringe — the septa used on intravenous canulae are ideal.

3.5.3 *Anaerobiosis*

It is essential to exclude oxygen rigorously and the measurements should be carried out under argon or nitrogen gas containing less than 1 p.p.m. O_2. All tubing should be made of an oxygen-impermeable material such as copper, Nylon 6 or Teflon. If necessary, Fieser's solution (20) can be used to remove residual oxygen from the gas. An apparatus of the 'Nilox Scrubber' type can be used — the details of this can be found in ref. 21, but they are also available commercially. After passage through Fieser's solution the gas should be cleaned in a dilute acid wash, a lead acetate wash and finally distilled water for pure gas. A recipe for Fieser's solution is given in *Table 6*. The gas inlet and outlet on the cuvette can conveniently be made from syringe needles and the attachments made by standard Luer fittings.

3.5.4 *Stirring*

Because of the depth of the column of liquid in a redox cuvette, adequate stirring can be a problem. The most efficient method is to stir from the top although this can be

Table 6. Fieser's Reagent[a].

1.	Add 2 g of anthraquinone-2-sulphonic acid (sodium salt) and 15 g of sodium dithionite to a warm 20% aqueous potassium hydroxide solution.
2.	Stir until they dissolve to give a blood red solution. Cool.
3.	100 ml will take 788 ml oxygen.
4.	Requires a dilute acid wash, a lead acetate wash (for hydrogen sulphide) and a distilled water wash.

[a]See ref. 20.

difficult to accommodate with commercially available electrodes because of space considerations. Dutton (19) describes an apparatus where a magnetic flea is used but is held to the side of the cuvette. Modern magnetic stirrers can be fitted into the base of the cuvette housing and the solution stirred from the bottom.

3.5.5 *Mediators*

Proteins, whether membrane-bound or soluble, are not able to interact directly with a platinum electrode. It is necessary to add small molecules which are able to interact both with the electrode and with the redox centres in the proteins to bring the two into equilibrium, and to ensure that the redox potential experienced by the protein is the same as that measured by the electrode; these substances are known as mediators.

Mediators are themselves redox couples and function efficiently only near to their own mid-point potential. Therefore a range of mediators is needed to cover the range of potentials to be titrated. Mediators should be stable, and for spectrophotometric titrations should be uncoloured in the region of interest. Unfortunately mediators generally have at least one highly coloured form and it is important to check that there is no spectral interference with the titration. Titrations should be independent of the mediator concentration used, and to check that mediators do not interfere, the titration should give the same results if the mediator concentration is changed, say by a factor of 5 or 10. If membrane-bound components are to be titrated, the mediators should be hydrophobic enough to interact with them. These ideals are frequently difficult to meet and it is often necessary to compromise on the mediators actually used.

Most mediators are unstable to light or air (or both) over periods of days, and ideally should be made up fresh for each day's experiments. A list of useful mediators, together with their mid-point potentials, is given in *Table 7*.

Mediators are selected bearing in mind that they work most effectively near their mid-point potential, say 60 mV either side of the mid-point potential if $n = 1$. Mediators should be present to cover effectively the whole range of the titration. For example, for a titration from $E_h = +100$ mV to $E_h = -50$ mV a mixture of N-methyl phenazonium methosulphate (PMS), N-ethyl phenazonium ethosulphate (PES), duroquinone, pyocyanine and 5-hydroxl-1:4-naphthoquinone should give adequate mediation. Generally, there is no harm in including more mediators covering the correct range than the absolute minimum; moreover the mediation should be more effective. There are regions of potential where few good mediators are available at the moment and extra care is needed to ensure full equilibration between electrode and sample. Care should be taken to avoid using a mediator which absorbs strongly at the wavelengths of interest. For example, titrations of α-bands of cytochromes cannot be done with TMPD because

Table 7. Mediators.

Name	E_{m7}	*n*	*Typical concentration*	*Dissolve in*
Potassium ferricyanide[a]	+430 mV	1	7.5 μg ml^{-1}	Aqueous
Hydroquinone	+280 mV	2	2.5 μg ml^{-1}	Ethanol
TMPD (N,N,N′,N′-tetramethyl phenylene diamine)	+260 mV	1	3 μg ml^{-1}	Ethanol
DAD (2,3,5,6-tetramethyl phenylene diamine)	+220 mV	2	12 μg ml^{-1}	Ethanol
PMS (N-methyl phenazonium methosulphate)	+80 mV	2	7 μg ml^{-1}	Aqueous
PES (N-ethyl phenazonium ethosulphate)	+55 mV	2	7 μg ml^{-1}	Aqueous
5-Hydroxyl-1,4-naphthoquinone	+50 mV	2	3 μg ml^{-1}	Ethanol
Duroquinone	+10 mV	2	4 μg ml^{-1}	Ethanol
Pyocyanine (N-methyl-1-hydroxy phenazonium methosulphate)	−34 mV	2	1.6 μg ml^{-1}	Aqueous
Anthraquinone	−100 mV	2	1 in 500 dilution of saturated solution	Ethanol
2-Hydroxyl-1,4-naphthoquinone	−145 mV	2	3 μg ml^{-1}	Ethanol
Anthraquinone-2,6-disulphonate	−185 mV	2	7.5 μg ml^{-1}	Aqueous
Anthraquinone-2-sulphonate	−225 mV	2	7.5 μg ml^{-1}	Aqueous
Benzyl viologen (1,1′-dibenzyl-4,4′-bipyridylium dichloride)	−311 mV	1	1 μg ml^{-1}	Aqueous
Methyl viologen (1,1′-dimethyl-4,4′-bipyridylium dichloride)	−430 mV	1	1 μg ml^{-1}	Aqueous

[a]See ref. 22.

of such an overlap of absorbance. Some mediators have specific interactions with particular components which may be undesirable: for example, neutral red is reported to interact specifically with the low potential electron acceptor, Q_L, of PSII, and many quinones quench chlorophyll fluorescence. Published titrations will give an idea of the mediators to use for any particular component.

The concentrations of mediators generally used are in the range 10−50 μM. For redox poising experiments, where flash-induced kinetics are being followed, it may be necessary to use lower concentrations of mediator than for equilibrium titrations.

It is important that the interaction of the mediators with the redox chain is slow compared with the change being measured. Some mediators such as PMS can act to bypass electron transport components and themselves reduce other members of the chain; if the mediator is present in too high a concentration the titration will be distorted. In any case, the titration should always be checked by repeating with a different concentration of mediators present.

3.5.6 *Testing*

When a new redox apparatus is put into use it can be tested by titrating known components. One of the simplest is the ferri/ferrocyanide couple, which does not require the use of any additional instrumentation. Values for the mid-point potential of the ferri/

ferrocyanide couple are given in ref. 22. If a spectrophotometer is available a mediator can be titrated and followed optically, or more biologically, cytochrome *c* can be used; information on the mid-point potential for cytochrome *c* is given in ref. 23.

If difficulty is experienced in maintaining low potentials, this may indicate that oxygen levels are too high; either the gas contains too much oxygen, or there is a leak in the system. It is essential that stirring is maintained at an adequate rate; if additions of reductant or oxidant take a long time to alter the measured potential it may be that stirring is insufficiently vigorous.

3.6 **Performing Redox Titrations**

The precise details of the technique will depend on the type of measurements involved and on the instrumentation available. The description below is given as a guide and refers to a titration of a membrane preparation of the photosynthetic bacterium *Rhodopseudomonas capsulata* Str. AlAPho$^+$. This strain is defective in carotenoid synthesis and makes no carotenoids with significant absorbance at wavelengths where the α-bands of cytochromes absorb, permitting the redox state of *b*- and *c*-type cytochromes to be monitored easily. In this case a *c*-type cytochrome has been titrated by following the absorbance at 551−540 nm.

Because of light scattering problems and absorbance of pigments, it is often necessary to limit the concentration of the species to be titrated. It is also necessary, however, to have sufficient absorbance change between the totally oxidised and the totally reduced forms of each species present to be able to measure possibly as many as 15 or 20 different points. For most purposes the total reduced-minus-oxidised absorbance can conveniently be measured using potassium ferricyanide to oxidise the sample and sodium dithionite to reduce it, these reagents having potentials beyond the ranges of most components involved in photosynthetic electron flow.

(i) To estimate the concentration of sample required in the titration, suspend the membranes in the buffer to be used [e.g. 50 mM 3-(N-morpholino)propanesulphonic acid (MOPS), 50 mM KCl, pH 7.0] to about the right concentration — 150 μM bacteriochlorophyll (BChl) under the above conditions.

(ii) Oxidise the reference sample completely with a few crystals of potassium ferricyanide.

(iii) Add a little solid sodium dithionite to reduce the sample (solid dithionite is easily oxidised on exposure to air so the bottle must be kept sealed and the surface layer of dithionite avoided). Leave for a couple of minutes and record the spectrum.

(iv) Check that the sample is totally reduced by adding a little more dithionite and recording another spectrum. From the total reduced minus oxidised absorbance, the optimum concentration to use can easily be calculated, typically 100−200 μM BChl. An excess of dithionite may cause the membranes to precipitate. In the absence of mediators, some components, particularly membrane-bound components, are only slowly reduced by dithionite because of kinetic limitations.

Having established the optimum BChl concentration, the titration can now be performed.

(i) Dilute the sample in buffer that has been thoroughly bubbled with argon (or nitrogen) to remove the oxygen. Transfer to the redox cuvette and, passing argon over the solution continuously, stir for about 20 min. If a reference cuvette is required, enough sample should be diluted to be split between reference and sample cuvettes to ensure a good optical match between the two. Argon needs only to be bubbled through slowly once the apparatus is set up, one bubble every couple of seconds is enough to maintain anaerobicity. This can be monitored by bubbling the outlet gas from the cuvette through water. At this stage ensure there is no gas leak from the system.

(ii) Add the mediators through the septum. The choice of suitable mediators is discussed above. DAD and TMPD are slowly destroyed at high potential so should only be added when the potential is below +300 mV. Ferricyanide was the only mediator present in the titration illustrated in *Figure 12*.

(iii) If a reference cuvette is being used, it should be brought to the desired potential. Since that absorbance has to be constant over the whole course of the titration, which may be 2 h or so, the poise of the redox components has to remain steady for that length of time. The simplest approach is to oxidise them completely by addition of a crystal or two of ferricyanide.

(iv) Adjust the potential of the redox cuvette to the starting point for the titration. Make up a range of concentrations of sodium dithionite and potassium ferricyanide. As a starting solution, dissolve enough to cover the end of a microspatula in 1 ml or so of buffer.

(v) Add aliquots of 1 or 2 μl of dithionite through the septum to decrease the potential, and of ferricyanide to increase the potential. The amount that it is necessary to add to achieve the desired change in potential must be established by trial and error. Many samples, such as bacterial chromatophores, will contain endogenous reductants which act to bring the redox potential down during some parts of the titration, without the addition of external reductant. This can be exploited, if the rate of drift of the potential is compatible with full equilibration, as a convenient way to decrease the potential.

(vi) Allow the sample to remain at the starting potential for long enough for full equilibration. When the sample is equilibrated there should be no change in the absorbance while there is no change in the potential. The time required to reach equilibration will depend on how far the potential has been changed; at the initial point of the titration, which may mean a relatively large change of potential, at least 10 min should be allowed for the sample to equilibrate. If the potential is unstable and drifts away significantly from the desired potential it must be brought back and re-equilibrated. If the potential does not change on addition of dithionite or ferricyanide, check that the solution is being stirred efficiently; if it is, then try with a stronger solution of dithionite or ferricyanide.

Unbuffered dithionite solution is not stable at room temperature, in air, and must be made up in buffer and left with a stream of argon or nitrogen bubbling through. It is more stable in basic solution than in acidic solution. It is generally obvious when the dithionite solution has 'gone off' since it is no longer effective in bringing the potential down and begins to accumulate a dirty white precipitate. Potassium ferricyanide solution is stable for 1 day and should then be discarded.

(vii) Scan a spectrum over the desired range of wavelengths, which will include a wavelength to measure the change and an isosbestic wavelength as a reference (Chapter 4).

(viii) Change the potential by about 9–10 mV by addition of aliquots of dithionite or ferricyanide. Allow to equilibrate for about 4 min at the new potential (equilibration of membrane-bound components is generally more difficult than that of soluble components) and record another spectrum. Continue to the desired end point or until there is no longer any change. If titrating several components over a wide range of potential it is better to do more than one titration, over different parts of the range, overlapping to allow them to be matched up. The presence of endogenous reductants and differences in redox buffering through the mediators can lead to it being easier to poise the potential in some regions than in others.

(ix) At the end point of the titration, take the potential back up to the starting point and record the spectrum again to check that it has not altered. One or two additional points of the titration can also be checked, making sure that after each change of the potential the sample is left sufficiently long to equilibrate.

(x) Repeat the titration using a fresh sample but this time changing the potential in the opposite direction. There should be no hysteresis between the two titration curves; if there is, it indicates that the sample was not fully equilibrated. Try leaving it a longer time at the potential before taking a reading at each point and check that the correct mediators for that potential range are present in the correct concentrations.

3.7 Treatment of Data

(i) At each potential, record the difference in absorbance between two suitable wavelengths — a measuring wavelength and a wavelength which is isosbestic for the absorbance change of the component of interest. Plotting the value of the absorbance difference against ambient potential will give a graph of the type shown in *Figure 11*. *Figure 11a* represents a redox couple where one electron is involved ($n = 1$), *Figure 11b* a redox couple where two electrons are involved. However, this type of plot is less useful for equilibrium titrations than the one discussed below.

(ii) To get a straight line graph, from the relationship above,

$$E_h = E_m + \frac{59}{n} \log_{10} \frac{\text{oxidised}}{\text{reduced}}$$

where $2.3 \frac{RT}{F} = 59$ mV at 25°C and n is the number of electrons involved in the reduction, plot $\log_{10} \frac{\text{oxidised}}{\text{reduced}}$ against E_h.

(iii) Calculate a value for the amount of oxidised component at each potential E_h, by taking the difference between the absorbance of the totally reduced component and the absorbance of the reduced component at E_h, for all potentials.

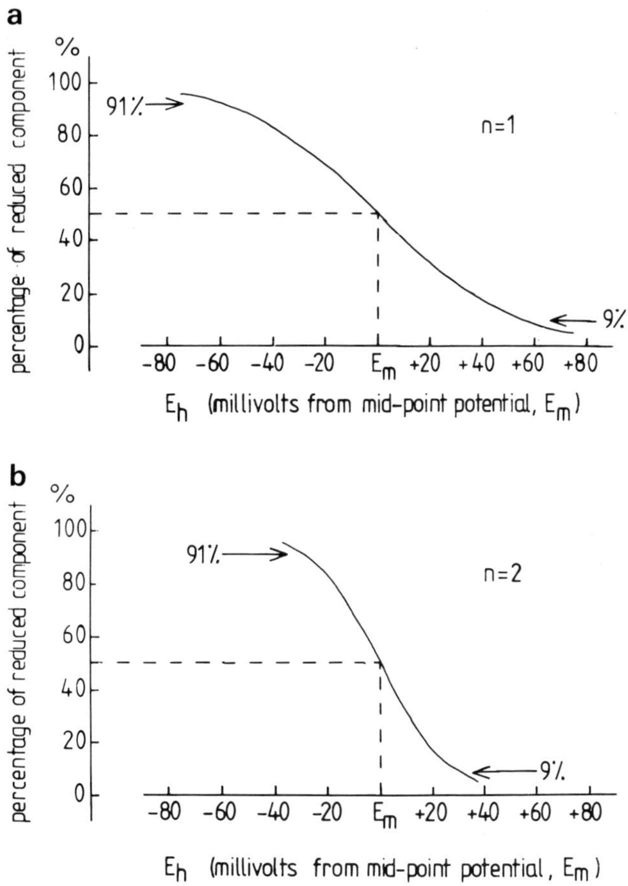

Figure 11. The dependence of the proportion of reduced species on redox potential for **(a)** a one-electron ($n = 1$) reaction and **(b)** a two-electron ($n = 2$) reaction.

(iv) Calculate the ratio of oxidised to reduced at each potential. Plot the logarithm of this ratio against ambient potential E_h, to get a graph of the sort shown in *Figure 12*.

If an $n = 1$ couple is titrated, a slope of 59 mV/decade change should be found, for an $n = 2$ couple the slope is 29.5 mV/decade change at 25°C. If a straight line is not obtained it may indicate a lack of equilibration between measurements (Section 3.6). The mid-point potential can then be read from the graph at the point where the ratio of oxidised to reduced species is equal to 1, i.e. where $\log_{10}$ (oxidised/reduced) is equal to 0. This corresponds to the point of maximum slope on the plot shown in *Figure 11*.

In the example given, the mid-point potential of the *c*-type cytochrome is +344 mV and the slope of the line indicates a one-electron reaction.

It may be that the titration involves more than one component with similar spectral

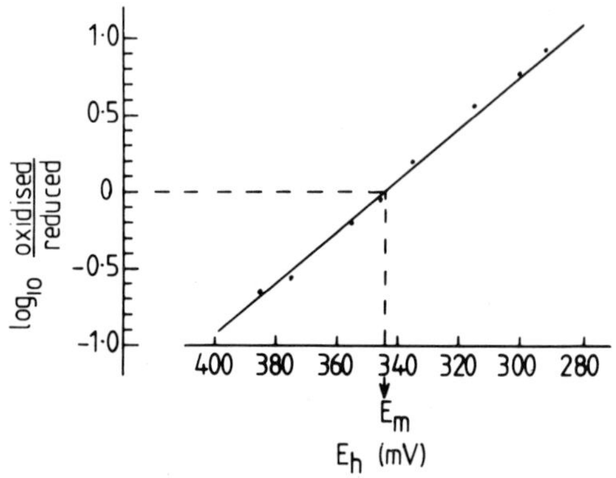

Figure 12. Redox titration of chromatophores of *Rhodopseudomonas capsulata* AlAPho$^+$, grown with N_2 as the sole nitrogen source, suspended to 150 μM BChl in 50 mM MOPS, 50 mM KCl, pH 7.0. 100 μM potassium ferricyanide was present as a mediator. The titration was conducted in the direction of decreasing potential. The measurements were made on a computer-linked single beam scanning spectrophotometer (N.G. Holmes and A.R.Crofts, unpublished observation).

characteristics, but with different mid-point potentials. As a rule of thumb, for an $n = 1$ plot, the change from 9% to 91% oxidised (or reduced) takes 118 mV (see *Figure 11*), similarly for an $n = 2$ plot, 9% to 91% takes 59 mV; if changes occur over a much wider span, then this suggests the titration of more than one component. It may be necessary in such multi-component titrations to try a number of plots, varying the parameters, and seeing which gives the optimal fit. This is most simply done by computer fitting of the data. In any case, if the components are close together then it becomes impossible to resolve them by eye and a computer must be used.

If the components are not too close together, and a clear plateau is visible between the two regions of changing absorbance with potential, they can be calculated as though they were two separate titrations.

4. ACKNOWLEDGEMENTS

We would like to thank the many colleagues who have contributed unwittingly to this chapter through comments and discussions over a number of years. We are indebted to Christine Sanders for artwork and for help with preparation of the manuscript, and thank Alan Logan for photography.

5. REFERENCES

1. Hill,R. and Bendall,F. (1960) *Nature,* **186**, 136.
2. Wraight,C.A. (1982) in *Photosynthesis Vol. 1. Energy Conservation by Plants and Bacteria,* Govindjee (ed.), Academic Press, London and New York, p. 17.
3. Danks,S.M., Evans,E.H. and Whittaker,P.A. (1983) *Photosynthetic Systems: Structure, Function and Assembly,* Wiley, New York.
4. Govindjee (1982) *Photosynthesis Vol. 1, Energy Conservation by Plants and Bacteria,* Academic Press, London and New York.
5. Umbreit,W.W., Burris,R.H. and Stauffer,J.F. (1972) *Manometric Techniques,* 5th edition, Burgers Publishing Co., Minneapolis.
6. Radmer,R. and Ollinger,O. (1980) in *Methods in Enzymology,* Vol. **69**, San Pietro,A. (ed.), Academic Press, London and New York, p. 547.
7. Bults,G., Horwitz,B.A., Malkin,S. and Cahen,D. (1982) *Biochim. Biophys. Acta,* **679**, 452.

8. Joliot,P. (1972) in *Methods in Enzymology*, Vol. **24**, San Pietro,A. (ed.), Academic Press, London and New York, p. 123.
9. Delieu,T. and Walker,D.A. (1972) *New Phytol.*, **71**, 201.
10. Allen,J.F. and Hall,D.O. (1974) *Biochem. Biophys Res. Commun.*, **58**, 579.
11. Trebst,A., Reimer,S. and Dallacker,F. (1976) *Plant Sci. Lett.*, **6**, 21.
12. Trebst,A. (1972) in *Methods in Enzymology*, Vol. **24**, San Pietro,A. (ed.), Academic Press, London and New York, p. 146.
13. Trebst,A. (1980) in *Methods in Enzymology*, Vol. **69**, San Pietro,A. (ed.), Academic Press, London and New York, p. 675.
14. Izawa,S. and Good,N.E. (1972) in *Methods in Enzymology*, Vol. **24**, San Pietro,A. (ed.), Academic Press, London and New York, p. 355.
15. Horton,P. (1983) *Proc. R. Soc. Lond. B*, **217**, 405.
16. Delieu,T.J. and Walker,D.A. (1983) *Plant Physiol.*, **73**, 534.
17. Nicholls,D. (1982) *Bioenergetics. An Introduction to the Chemiosmotic Theory*, Academic Press, London and New York.
18. Dutton,P.L. and Wilson,D.F. (1974) *Biochim. Biophys. Acta*, **346**, 165.
19. Dutton,P.L. (1978) in *Methods in Enzymology*, Vol. **54**, Fleischer,S. and Packer,L. (eds), Academic Press, London and New York, p. 411.
20. Fieser,L.H. (1924) *J. Am. Chem. Soc.*, **46**, 2639.
21. Gilroy,D. and Mayne,J.E.O. (1962) *J. Appl. Chem.*, **12**, 382.
22. Reilly,J.E. (1973) *Biochim. Biophys. Acta*, **292**, 509.
23. Henderson,R.W. and Morton,T.C. (1968) in *CRC Handbook of Biochemistry. Selected Data for Molecular Biology*, Soker,H.A. (ed.), Chemical Rubber Co., Cleveland, Ohio, J-35.
24. Allen,J.F. and Horton,P. (1981) *Biochim. Biophys. Acta.*, **638**, 290.

APPENDIX

Suppliers

Cuvette	York Glassware Services, 9 The Crescent, Blossom Street, York Y02 2AW, UK
Redox electrode	Kent Industrial Measurements, EIL Analytical Instruments, Hanworth Lane, Chertsey, Surrey, UK
	Russell pH Ltd., Station Road, Auchtermuchty, Fife KY14 7DP, UK
Septum and tubing	Portex Ltd., Hythe, Kent CT21 6JL, UK
Oxygen electrodes	Hansatech Ltd., Paxman Road, Hardwick Industrial Estate, King's Lynn, Norfolk, UK
	Rank Bros., High Street, Bottisham, Cambridge CB5 9DA, UK
Electronic stirrers	Rank Bros., High Street, Bottisham, Cambridge CB5 9DA, UK
Fibre-optic light sources	CUEL, 12 Tulip Tree Avenue, Kenilworth, Warwickshire CV8 2BU, UK
LED light sources	Hansatech Ltd., Paxman Road, Hardwick Industrial Estate, King's Lynn, Norfolk, UK
Photoacoustic sensor	Applied Photophysics, 20 Albemarle Street, London W1X 3HA, UK

CHAPTER 6

Photophosphorylation

J.D. MILLS

1. INTRODUCTION

Photophosphorylation is the overall process whereby the energy made available from absorbed light is used by the chloroplast to bring about phosphorylation of ADP. As indicated by *Figure 1*, this overall process is complex and requires the participation of the light-harvesting apparatus, reactions of electron transport and a reversible protonmotive ATPase, CF_0-CF_1. Energy coupling between electron transport and ATP synthesis is believed to be chemiosmotic (1,2), but although accepted in outline, some of the details of the process depicted in *Figure 1* are either disputed or not known. In contrast to this theoretical complexity, photophosphorylation and many of its associated partial reactions are in practice quite simple to measure in the laboratory.

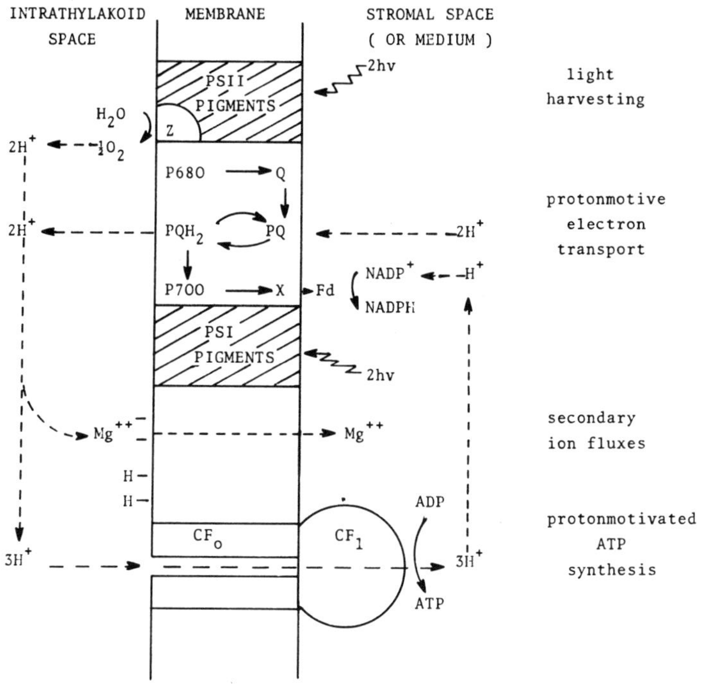

Figure 1. Schematic representation of chloroplast thylakoid photophosphorylation. PS, photosystem; Z, water-splitting enzyme; P680, P700, reaction centre chlorophylls; Q, X, primary electron acceptors; PQ, plastoquinone; Fd, ferredoxin; hν, light.

Often little in the way of sophisticated or expensive apparatus is required. This chapter will concentrate on presenting methods that require the minimum of specialised equipment yet which can be relied upon to give reproducible and quantitative results. In keeping with the theme of this book, the aim is to provide a simple method that can be rapidly set up in a laboratory with little prior experience in the field. However, for each technique, hints are also given on how to upgrade the method so as to provide increased sensitivity or sophistication suitable for prolonged or intensive exploitation of the technique.

2. PREPARATION OF CHLOROPLASTS

In general, neither the method nor source material for chloroplast preparation is critical for these techniques provided that well coupled thylakoids can be obtained. However, some words are necessary on the use of either intact or lysed chloroplasts as the stock suspension. There are certain advantages in preparing intact chloroplasts and maintaining these as the stock suspension. The chief advantage is the low rate of general deterioration of the organelles after preparation. Since most of the techniques require lysed organelles, lysis can be done *in situ* just prior to experimentation. If spinach or pea seedlings are used as the source material, then intact chloroplasts should be prepared as described in Chapter 2 and resuspended in resuspension medium RM1 (*Table 1*). Avoid using grinding media containing phosphate buffers if experiments on ATP synthesis or hydrolysis are to be performed.

For certain techniques however, it is necessary to prepare thylakoids that have been lysed and washed during the preparation. The following method should be used and is also suitable if for any reason intact chloroplasts cannot be prepared.

2.1 **Preparation of Washed, Lysed Chloroplasts**

(i) Prepare chloroplasts as if intact according to the methods in Chapter 2.
(ii) Lyse the chloroplasts by resuspending in 10 ml of chilled lysis medium LM1 (*Table 1*) and store on ice for 60 sec.
(iii) Add 40 ml of chilled resuspension medium RM2 (*Table 1*) and collect the thylakoids by centrifugation at 3000 *g* for 5 min.
(iv) Resuspend the thylakoids in RM1 unless a low buffering medium is required, in which case use RM2.

Table 1. Media for the Preparation and Storage of Chloroplast Stock Suspensions.

RM1	*LM1*	*RM2*
0.33 M sorbitol	20 mM KCl	0.35 M sorbitol
2 mM EDTA	5 mM $MgCl_2$	5 mM ascorbic acid
1 mM $MgCl_2$	10 mM MES-KOH	20 mM KCl
1 mM $MnCl_2$	pH 6.5	5 mM $MgCl_2$
50 mM Hepes-KOH		10 mM MES-KOH
pH 7.6		pH 6.5

RM2 should be prepared just prior to use.

(v) Store the thylakoids on ice.

All chloroplast stock solutions will be assumed to be resuspended at a chlorophyll concentration of approximately 2 mg chlorophyll ml^{-1}.

2.2 Preparation of Thiol-modulated Thylakoids

2.2.1 *Applications and Advantages*

CF_0-CF_1 can exist in two forms depending on the redox state of its thiol groups (see Section 4.1 for details). Only the thiol-reduced form is observed to catalyse net ATP hydrolysis and so the enzyme must be brought into this form when ATP hydrolysis is to be measured. The interconversion of CF_0-CF_1 between its oxidised and reduced forms is a physiological process that is catalysed *in vivo* by the thioredoxin system (3). Thioredoxin *in vivo* is oxidised in the dark but reduced in the light by the action of photosynthetic electron transport. *In vivo* therefore, CF_0-CF_1 is thought to be in the oxidised form during darkness but in the reduced form in the light. Thylakoids in which CF_0-CF_1 is predominantly thiol-reduced may therefore be prepared by rapid isolation of broken chloroplasts from pre-illuminated leaves or intact chloroplasts. Conversely, when thylakoids with a low catalytic potential for ATP hydrolysis are required, care must be taken during the preparation to ensure that thiol modulation does not occur. This means that leaves must be kept in the dark prior to and during thylakoid preparation and intact chloroplasts must be stored on ice in the dark.

The method for inducing thiol modulation of CF_0-CF_1 given here does not depend on the ability to isolate intact organelles, or isolate thylakoids rapidly from pre-illuminated leaves. Instead, thylakoids are illuminated in the presence of dithiothreitol (DTT), a non-physiological reductant that induces complete conversion of the enzyme to its reduced form. The illumination is required to generate ΔpH, causing activation of CF_0-CF_1 to occur (see Section 4.1). DTT interacts much faster with CF_0-CF_1 when the enzyme is activated.

2.2.2 *Equipment*

A 250 W projector is used as the light source. The beam should be filtered through a Balzers Calflex C (or other heat-absorbing filter) together with a Corning 4-96 (green) or 2-62 (red) filter. Light can be focussed onto a reaction vessel using a 100 ml round-bottomed flask. If suitable glass filters are not available, fill the flask with a 1% $CuSO_4$ solution to provide blue-green light. The intensity of the light must be adjusted to 30–70 W m^{-2}. Use Kodak Wratten 96 neutral density filters mounted in 50 × 50 mm slide holders. A single 50% reduction in light intensity can be achieved by placing a 50 mm square piece of perforated zinc in the slide holder of the projector.

The reaction cuvette for this and many other techniques in this chapter should be a stirred, thermostatted vessel of 1–5 ml capacity that can be easily illuminated. Well-type oxygen electrodes (Hansatech Ltd., Paxman Rd, Kings Lynn, Norfolk, UK or Rank Bros., Bottisham, Cambridge, UK) are ideal for this purpose. Leave the electrode chamber open to the air.

It is convenient to carry out the reaction in plastic disposable test tubes, which can be inserted into the well of the O_2 electrode. Most 16 mm tubes will fit into the Rank electrode whilst 10 mm tubes are required for the Hansatech electrode. The advantage

Table 2. Media for Thiol-modulation of CF_0-CF_1.

LM2	
6 mM $MgCl_2$ 24 mM KCl 10 mM dithiothreitol 0.12 mM methyl viologen 750 units catalase ml^{-1} 30 mM Tricine-KOH pH 8.0	2 M sorbitol

Catalase (type C40) is obtained from Sigma Co. Prepare LM2 just prior to use.

of using tubes is that the reaction can be carried out under defined conditions of temperature and light intensity yet the tube and its contents can be rapidly removed (e.g. to store the treated organelles on ice). A second advantage is that the contents of the tube are isolated from the electrode, preventing contamination problems. Stirring is achieved by placing 6 mm stirring bars (Bel-Art Products, Pequannock, New Jersey 07440, USA) in the tubes.

2.2.3 *Method*

(i) Take 0.15 ml of intact or lysed chloroplasts (~0.3 mg chlorophyll) and add to 0.85 ml of lysis medium LM2 (*Table 2*) in the reaction vessel.

(ii) After 30 sec, add 0.2 ml of 2 M sorbitol to bring the final sorbitol concentration to 0.33 M and chlorophyll concentration to 0.25 mg ml^{-1}. Higher concentrations of chloroplasts can be used by varying the ratio of stock suspension to LM2 up to 1:4. Addition of sorbitol is not necessary, especially at higher chlorophyll concentrations.

(iii) Illuminate the suspension for 6 min at an intensity of no more than 70 W m^{-2}. The temperature should be 16–20°C.

(iv) If the treated thylakoids are not to be used immediately, store on ice and use within 30 min.

The conversion of CF_0-CF_1 to the thiol-reduced form may be followed by taking aliquots of the chloroplast suspension and assaying for ATPase activity as described in Section 4.2.1. *Figure 2* shows the rate of appearance of ATPase activity as a function of the illumination time in the presence of DTT. ATPase activity is slow to appear, reflecting the relatively slow rate of conversion of CF_0-CF_1 to the thiol-reduced form.

2.2.4 *Trouble-shooting*

The above method is reliable using chloroplasts prepared from peas or spinach grown at temperatures around 20°C or below (see autumn-grown spinach, *Figure 2*). With less perfect starting material, a general inhibition of photosynthesis is often induced by the long pre-illumination procedure given above (see summer-grown peas, *Figure 2*). This photoinhibition can be reduced by shortening the pre-illumination time, reducing the light intensity of 30 W m^{-2} and reducing the temperature to 16°C. However, all these modifications slow down the rate at which thiol modulation of CF_0-CF_1 occurs

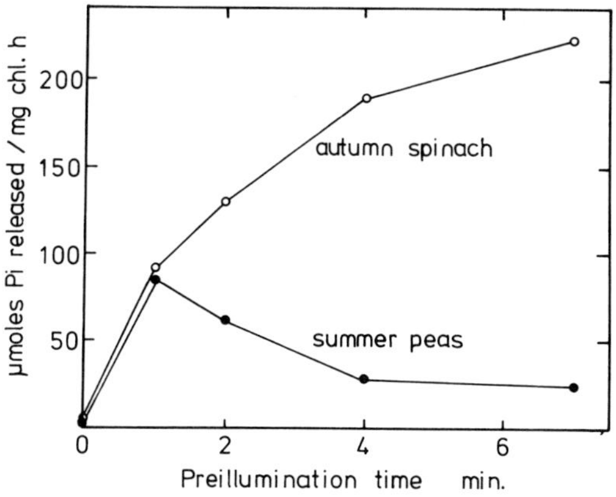

Figure 2. Induction of ATPase activity in thylakoids pre-illuminated in the presence of dithiothreitol. Thylakoids were prepared by lysing intact chloroplasts isolated from either spinach grown at 15 – 20°C during the autumn (open circles) or peas grown at >30°C during the summer.

and may cause the thiol modulation process to be incomplete on terminating pre-illumination. Increasing the concentration of DTT does not generally improve the rate of thiol modulation but can increase the extent of photoinhibition.

3. GENERATION AND MEASUREMENT OF ΔpH AND Δψ

Energy coupling between the reactions of electron transport and ATP synthesis is now generally agreed to occur according to chemiosmotic principles as suggested by Peter Mitchell (1,2). Before the methods of this section can be described, it is necessary to discuss the principles of energy coupling upon which these methods are based.

3.1 Principles

3.1.1 *Coupling of Electron Transport to ATP Synthesis*

According to chemiosmotic principles, useful work is obtained from the large free energy decrease of electron transport by coupling the latter to the translocation of protons against an electrochemical potential difference of protons ($\Delta\bar{\mu}_{H^+}$) across the thylakoid membrane. As shown in *Figure 1*, the reactions of non-cyclic electron transport are thought to be vectorially arranged in the thylakoid membrane so that the passage of one electron from H_2O to $NADP^+$ results in the net translocation of at least $2H^+$ from the stromal side of the thylakoid into the intrathylakoid space. (This H^+/e ratio may be >2 if electron transfer in the plastoquinone/cyt-*f* region of the chain occurs *via* a Q-cycle type mechanism, Chapter 1.) Cyclic electron transfer around photosystem I also results in the net translocation of protons into the thylakoid (*Figure 3*). When artificial cofactors such as phenazine methosulphate or diaminodurene (DAD, *Figure 3*) are used to catalyse cyclic electron transfer, the H^+/e ratio of cyclic electron flow is 1. (Again, this ratio may be >1 during endogenous ferredoxin-catalysed cyclic electron transport if a Q-cycle operates.) The net inward H^+ flux generated by both non-cyclic and cyclic electron flow is electrogenic, because the transfer of positively charged H^+ ions across the thylakoid translocates net charge across the membrane.

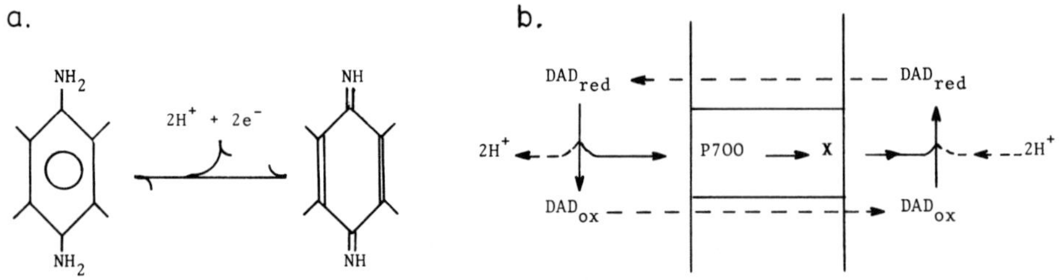

Figure 3. **(a)** Structure of oxidised and reduced diaminodurene. **(b)** Schematic representation of protonmotive cyclic electron transport around photosystem I catalysed by diaminodurene. Electron transfer events are represented by solid lines; diffusion of H^+ and diaminodurene by dotted lines. Note that other carriers (not shown) may be involved in electron transfer.

The thylakoid membrane of freshly prepared chloroplasts presents a barrier to the passive diffusion of H^+ ions. Thus electron transfer results in the accumulation of H^+ in the intrathylakoid space, resulting in a decrease in pH and an increase in electric potential relative to the outer bathing medium.

The translocation of a proton into the intrathylakoid space therefore increases its electrochemical potential, since the electrochemical potential of any ion, i ($\bar{\mu}_i$) depends on its valency (Z_i), activity ($a_i[C_i]$, where a_i is the activity coefficient) and electric potential (ϕ) of the phase in which it is dissolved (1).

$$\bar{\mu}_i = \mu_i^0 + Z_iF\phi + RT\ln(a_i[C_i]) \qquad \text{Equation 1}$$

where μ_i^0 is the standard chemical potential and F is the Faraday. The activity of H^+ ions can be well approximated by their concentration, and the electrochemical potential of H^+ on the inner and outer sides of the thylakoid can be defined by Equations 2 and 3.

$$\bar{\mu}_{H^+in} = \mu_{H^+}{}^0 + F\phi_{in} + RT\ln[H^+]_{in} \qquad \text{Equation 2}$$

$$\bar{\mu}_{H^+out} = \mu_{H^+}{}^0 + F\phi_{out} + RT\ln[H^+]_{out} \qquad \text{Equation 3}$$

Thus the difference in electrochemical potential, $\Delta\bar{\mu}_{H^+}$, between protons in the intrathylakoid space and outer aqueous phase is given by:

$$\Delta\bar{\mu}_{H^+} = \bar{\mu}_{H^+in} - \bar{\mu}_{H^+out}$$

$$= F\Delta\psi + RT\ln([H^+]_{in}/[H^+]_{out}) \qquad \text{Equation 4}$$

where the symbol $\Delta\psi$ represents the difference in electric potentials ($\Delta\phi$) between the two aqueous phases.

The right-hand expression of Equation 4 represents the purely chemical (concentration) term of $\Delta\mu_{H^+}$ and can be written in terms of pH, thus:

$$\Delta\bar{\mu}_{H^+} = F\Delta\psi - 2.303RT\Delta pH \qquad \text{Equation 5}$$

$\Delta\bar{\mu}_{H^+}$ has units of kJ mol^{-1} and represents the minimum work that has to be done by the reactions of electron transport when 1 mole of protons is translocated from the outer aqueous phase into the intrathylakoid aqueous space.

To facilitate comparisons with redox potential spans in the electron transport chain, Mitchell defined $\Delta\bar{\mu}_{H^+}$ in terms of electrical units which he called the 'protonmotive force', Δp (1).

$$\Delta p = \Delta\bar{\mu}_{H^+}/F$$

$$= \Delta\psi - 2.303(RT/F)\Delta pH \qquad \text{Equation 6}$$

At 25°C, Equation 6 has the numerical value:

$$\Delta p = \Delta\psi - 59\Delta pH \qquad \text{Equation 7}$$

Δp has units of mV and is therefore a potential rather than a force. This point is well discussed in (4) but readers are advised to return to the original literature (1,2) for a comprehensive treatment. A consistent sign convention will be maintained throughout this chapter, where Δ is taken to mean the value in the intrathylakoid space *minus* the value in the surrounding aqueous phase (that is the stroma in intact organelles and the outer suspending medium for isolated thylakoids). This means that Δp, $\Delta\bar{\mu}_{H^+}$ and $\Delta\psi$ will normally be positive but ΔpH will have a negative value.

The potential energy in the electrochemical H^+ difference becomes available when protons pass from the intrathylakoid space back into the outer aqueous phase. $\Delta\bar{\mu}_{H^+}$ represents the maximum work that can be done per mole of proton efflux, provided that some means exists for coupling the efflux to some other endergonic process. When proton efflux occurs *via* the CF_0-CF_1 complexes, then work is done and free energy is conserved by coupled ATP synthesis (phosphorylating H^+ efflux). If, however, H^+ efflux occurs by passive diffusion through the membrane (non-phosphorylating H^+ efflux), then no useful work is done, and the free energy originally conserved in the electrochemical proton gradient appears mainly as an increase in the entropy of the system. In freshly prepared thylakoids illuminated in the presence of ADP + Pi, most the H^+ efflux occurs *via* CF_0-CF_1 and ATP synthesis is said to be coupled to electron transport. When thylakoids are damaged during preparation so that the passive permeability of the membrane to H^+ is greatly increased, then non-phosphorylating H^+ efflux becomes significant and the capability for ATP synthesis is diminished. In this condition, the chloroplasts are said to be partially (or completely) uncoupled, as discussed further in Section 3.1.4.

3.1.2 *Steady-state and Pre-steady State Contributions of ΔpH and $\Delta\psi$ to $\Delta\bar{\mu}_{H^+}$*

It has been shown that a single turnover of the electron transport chain (induced by a short flash of light) creates a protonmotive force (Δp) of 53 mV, consisting mainly of $\Delta\psi$ (50 mV) and a very small contribution from the ΔpH component (-0.05 units $\equiv \Delta p = 3$ mV) (5). Thus the first protons to be electrogenically pumped into the thylakoid generate $\Delta\psi$ much faster than ΔpH. This arises because the electrical capacitance of the thylakoid is much smaller than the pH buffering capacity.

From the above data it is clear that very few turnovers of electron transport are required to generate Δp to its steady-state value of typically 180 mV. Indeed, ATP synthesis is observed to reach its maximum rate within the first second of continuous illumination. During this time, $\Delta\psi$ would be expected to be a major contributor to Δp and this has been observed (6). However, after 60 sec of continuous illumination, Δp consists almost entirely of ΔpH, and $\Delta\psi$ is observed to be rather small (6,7). Thus the first few seconds of illumination are characterised by a change in the relative energetic contributions of $\Delta\psi$ and ΔpH to Δp, and this arises because of secondary ion fluxes across the thylakoid membrane (8).

The thylakoid membrane is moderately permeable to both cations and anions other than H^+ and OH^-. The generation of $\Delta\psi$ causes electrophoretic movement of these ions, either as cation efflux, or as anion uptake into the thylakoid, depending on the composition of the suspending medium (8). These secondary ion fluxes during the first 60 sec of illumination cause the collapse of $\Delta\psi$ and allow continuous net uptake of H^+ ions, resulting in a large steady-state ΔpH being formed. However, it should be emphasised that although the ΔpH component thus requires 1 min to reach its steady-state value, Δp probably reaches its maximum well within the first second of illumination. It is only the relative energetic contributions of $\Delta\psi$ and ΔpH that vary over the first 60 sec or so of illumination.

3.1.3 *Profiles of $\Delta\psi$ and ΔpH Transverse to the Thylakoid Surface*

In the 'classical' view of chemiosmosis presented so far, the internal aqueous phase of the thylakoid has been assumed to be homogeneous and highly proton-conducting so that it is isopotential with respect to $\bar{\mu}_{H^+in}$. It should be noted here that there is a growing school of thought that the intrathylakoid space may contain subdomains where $\bar{\mu}_{H^+in}$ of one domain is not in equilibrium with other domains. However, since these domains have not yet been physically demonstrated, let alone methods devised for measuring $\bar{\mu}_{H^+in}$ within domains, this concept will not be discussed further here. Nevertheless, the reader should be aware that the assumption of a single, isopotential intrathylakoid phase is implicit in the design of most of the methods given and in the interpretation of their results.

Even though the intrathylakoid space is isopotential with respect to $\bar{\mu}_{H^+in}$, the energetic contributions of the electrical and pH components may vary with the distance from the thylakoid membrane surface. This effect arises because of the electrostatic properties of the thylakoid surface (9). Both surfaces of the thylakoid bear fixed negative charges which, on the inner surface, are responsible for the considerable pH buffering properties of the thylakoid. These fixed negative charges attract surrounding cations – including H^+ – to the surface forming a stable, diffuse ionic double layer, depicted schematically in *Figure 4a*. The net result is that the surface potential, (ψ_0) can be significantly lower than the electrical potential of the bulk solution (ψ), measured at an infinite distance from the surface (see *Figure 4b*). Similarly, the pH immediately adjacent to the surface (pH_0) may be lower than in the bulk solution at a distance from the surface as illustrated in *Figure 4c*. If the surface charge properties of the inner and outer thylakoid differ as shown in *Figure 4*, then the difference in surface potentials ($\Delta\psi_0$) may be larger than the $\Delta\psi$ between the bulk solutions. Correspondingly, the difference in pH between the two surfaces (ΔpH_0) will be less than the ΔpH between the two bulk solutions. However, $\Delta\bar{\mu}_{H^+}$ between the two surfaces will be the same as $\Delta\bar{\mu}_{H^+}$ between the two bulk solutions since each aqueous phase is isopotential with respect to its $\bar{\mu}_{H^+}$.

This theoretical complexity may give rise to practical difficulties when various probe molecules are used to measure ΔpH and $\Delta\psi$. For example, positively charged amines will tend to accumulate in the double-layer region of the intrathylakoid space and may give a distorted measure of ΔpH (which is too high) and reflects the surface pH as well as bulk pH of the intrathylakoid space. These effects can be minimised by adding 5 mM divalent cations which effectively screen the surface charges, reduce ψ_0 and decrease the thickness of the double layer (9). All the techniques described here use media containing 5 mM $MgCl_2$ to minimise these possible complications. All references

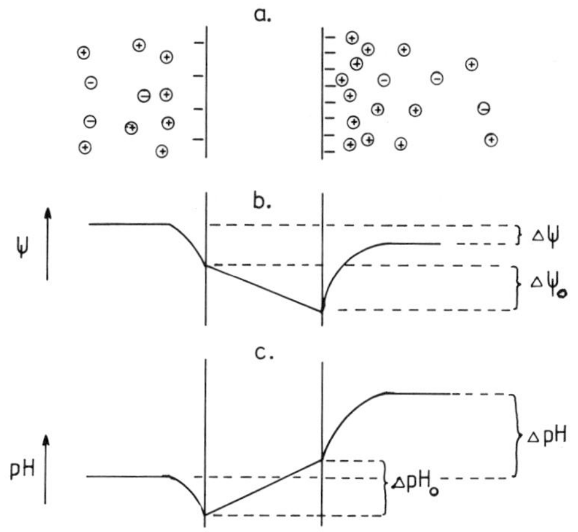

Figure 4. Influence of the electrical diffuse double layer on the profile of pH and ψ transverse to the thylakoid surface. **(a)** Schematic view of a diffuse double layer on either side of the membrane. The membrane is assumed to be negatively charged and the density of these charges is greater on the right-hand face of the membrane. **(b)** and **(c)** Profiles of pH and ψ transverse to the membrane and its surfaces. ΔpH and $\Delta\psi$ represent difference in pH and electrical potential between the bulk solutions; ΔpH_0 and $\Delta\psi_0$ the difference between the two surfaces of the membrane.

to ΔpH and $\Delta\psi$ outside of this section refer to the difference between bulk solutions as indicated in *Figure 4*.

3.1.4 *Uncouplers*

Uncouplers were originally defined by their ability simultaneously to abolish ATP synthesis and accelerate electron transport to its maximum rate. Within the framework of the chemiosmotic hypothesis, the coupling link between electron transport and ATP synthesis is thought to be $\Delta\bar{\mu}_{H^+}$, and uncouplers appear to operate by accelerating the dissipation of ΔpH, or $\Delta\psi$, or both, by accelerating non-phosphorylating ion fluxes across the thylakoid. The modern use of the term 'uncoupling' is now rather loose and tends to include any treatment that causes increased dissipation of ΔpH or $\Delta\psi$. Whilst this may lead to uncoupling in certain cases, the definition of uncoupling used here means the ability of the treatment significantly to decrease the capacity of the thylakoid to synthesise ATP under the given set of experimental conditions.

The mode of action of potentially uncoupling treatments and reagents (summarised in *Figure 5*) can be divided into several classes.

(i) Treatments that generally increase the overall permeability of the thylakoid to ions including H^+, e.g. detergents, membrane damage, etc.

(ii) Reagents that specifically increase the membrane conductance to H^+ ions, either by forming channels (e.g. gramicidin) or because both the protonated and unprotonated forms of the reagent are able to permeate across the thylakoid membrane (e.g. CCCP).

Both these types cause increased H^+ efflux uncompensated by other ion fluxes across the membrane and hence dissipate both ΔpH and $\Delta\psi$.

(iii) Weak bases that freely permeate the membrane in the unprotonated state but are

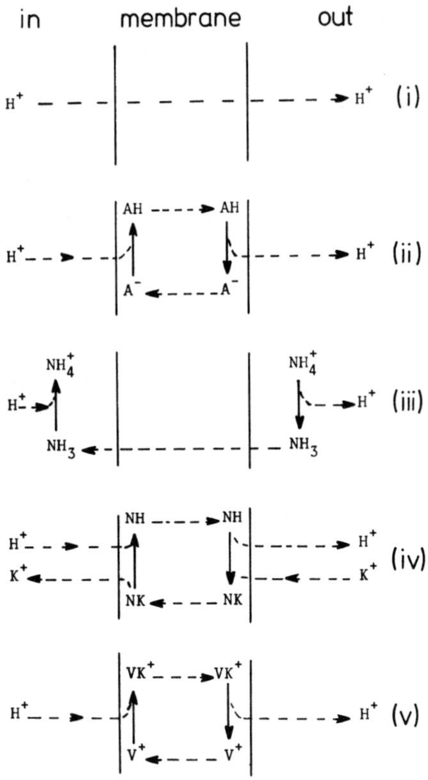

Figure 5. Schematic representation of the five modes of action of potentially uncoupling reagents. Electron transport would translocate protons from right to left across the membrane. See text for other details.

entrapped within the thylakoid upon protonation, e.g. amines. As indicated in *Figure 5(iii)*, this type of mechanism is electroneutral and effectively exchanges one H^+ for one $NH_4{}^+$ ion across the thylakoid.

(iv) Ionophores such as nigericin [*Figure 5(iv)*] that catalyse a strictly electroneutral exchange of H^+ ions for monovalent cations across the thylakoid. This mechanism depends on the ability of the ionophore to complex a cation of H^+ in a hydrophilic pocket but to remain lipophilic by virtue of the 'shell' of hydrophobic groups presented to the exterior.

Both type (iii) and (iv) mechanisms cause the dissipation of ΔpH without affecting $\Delta\psi$ since no net charge crosses the membrane.

(v) Ionophores such as valinomycin that specifically increase the membrane permeability to ions other than H^+. These reagents facilitate passive diffusion of the ion across the membrane and allow electrophoretic ion fluxes in response to any $\Delta\psi$, thereby collapsing $\Delta\psi$ without affecting ΔpH.

Whether or not the potentially uncoupling reagents of type (iii)–(v) are effective uncouplers will depend on the conditions of the experiment. Under steady-state conditions of continuous illumination, $\Delta\bar{\mu}_{H^+}$ is composed almost entirely of the ΔpH component. Those reagents that dissipate ΔpH are effective uncouplers of ATP synthesis whereas valinomycin (+ K^+), which dissipates $\Delta\psi$, is not. However, during single flashes of light, or during the first second of continuous illumination, when $F\Delta\psi$ is a major component of $\Delta\bar{\mu}_{H^+}$, valinomycin (+ K^+) is much more effective as an uncoupler of ATP synthesis (6).

The effectiveness of uncoupling will also depend on the concentrations of uncoupler and any co-ions involved in the uncoupling mechanism. In theory, full uncoupling during continuous illumination will only occur if the potential flux of H^+ into the thylakoid is matched by a similar flux of amine [type (iii)] or K^+ ions [type (iv)], so preventing the development of $\Delta\bar{\mu}_{H^+}$. Obviously, the concentration of nigericin must be high enough to catalyse the required K^+/H^+ exchange rate. The dependence on amine concentration (or K^+ in the presence of nigericin) is more subtle. Continuous uptake of NH_4^+ into the thylakoid generates an electrochemical potential difference of NH_4^+ ions ($\Delta\bar{\mu}_{NH_4^+}$) across the membrane:

$$\Delta\bar{\mu}_{NH_4^+} = F\Delta\psi + 2.303RT\log\left(\frac{a_{in}[NH_4^+]_{in}}{a_{out}[NH_4^+]_{out}}\right) \qquad \text{Equation 8}$$

Continuous uptake of NH_4^+ ions against $\Delta\bar{\mu}_{NH_4^+}$ can only occur if:

$$\Delta\bar{\mu}_{H^+} > \Delta\bar{\mu}_{NH_4^+} \qquad \text{Equation 9}$$

It is therefore important to realise that uncouplers of type (iii) [and (iv)] can never decrease $\Delta\bar{\mu}_{H^+}$ below that of the $\Delta\bar{\mu}_{NH_4^+}$ developed as a result of the uncoupling process. To ensure effective uncoupling, conditions must be chosen so that $\Delta\bar{\mu}_{NH_4^+}$ does not become significant. This is most simply achieved by maintaining an initial concentration of at least 10 mM NH_4Cl (or 10 mM KCl in the presence of nigericin). Although, in principle, continuous uptake of amine will eventually cause $\Delta\bar{\mu}_{NH_4^+}$ to become significant, in practice this does not occur for several reasons that are best illustrated by an example calculation.

Imagine ATP synthesis is being measured over the first minute of illumination in the presence of 10 mM NH_4Cl and an acceptor for non-cyclic electron transport. If the rate of electron transport is 600 μeq (mg chl)$^{-1}$ h^{-1} then the rate of H^+ influx will be 1200 μmol (mg chl)$^{-1}$ h^{-1} or 20 μmol H^+ (mg chl)$^{-1}$ min^{-1}. Assuming that for nearly every H^+ initially translocated, one NH_4^+ ion is taken up, then approximately 20 μmol of NH_4^+ will be taken up during the first minute of illumination. Assuming the internal volume of thylakoids to be 20 μl (mg chl)$^{-1}$ under these conditions, then the internal concentration of NH_4^+ will have risen to 1 M and $[NH_4^+]_{in}/[NH_4^+]_{out}$ will be 100. From Equation 8, $\Delta\bar{\mu}_{NH_4^+}/F$ will have the value of 120 mV, assuming activity coefficients are equal inside and outside the thylakoid.

Thus it can be seen from this calculation that Δp will have risen to a little over 120 mV during this period of the experiment, a value too small to drive significant ATP synthesis. In fact, $\Delta\bar{\mu}_{NH_4^+}/F$ will be less than 120 mV because at a concentration of 1 M inside and 10 mM outside, the activity coefficients for NH_4^+ ions will not be equivalent. Of more importance, the uptake of such large quantitites of NH_4^+ ions must be balanced by Cl^- uptake to prevent build-up of $\Delta\psi$. The net uptake of NH_4Cl will lead to osmotic water entry, extensive swelling of the thylakoids and eventually to uncoupling *via* type (i) process as the thylakoid membrane becomes damaged. Thus uncoupling in the presence of 10 mM NH_4Cl will be effective even though some $\Delta\bar{\mu}_{NH_4^+}$ will inevitably develop.

In fact, the development of $\Delta\bar{\mu}_{NH_4^+}$ can be put to good use in methods for measuring ΔpH. Let us repeat the above calculation but let the initial concentration of NH_4Cl be 15 μM instead of 10 mM and let the chlorophyll concentration in the experiment be 10 μg chlorophyll ml^{-1}. Again the uncoupling process will cause the initial rate of NH_4^+ uptake to be 20 μmol NH_4 (mg chl)$^{-1}$ min^{-1} and so in the first 3 sec, 10 nmol ml^{-1} will be taken up by the chloroplasts, depleting the medium to 5 μM NH_4Cl. The internal concentration of NH_4^+ will have risen to 50 mM, and $[NH_4^+]_{in}/]NH_4^+]_{out}$ will be 10^4. From Equation 8, again neglecting activity coefficients (this time with more justification), $\Delta\bar{\mu}_{NH_4^+}/F$ will be 240 mV and from Equation 9, $\Delta\bar{\mu}_{H^+}/F$, or Δp must be greater than 240 mV.

As discussed in Section 4.1, a value of $\Delta p = 240$ mV is sufficient to maintain ATP synthesis at its maximum rate. Uncoupling by 15 μM NH_4Cl will therefore only be transient and will be over within 3 sec of continuous illumination. After this time, net NH_4^+ fluxes will cease and the electrochemical potential differences of H^+ and NH_4^+ will come into equilibrium:

$$\Delta\bar{\mu}_{H^+} = \Delta\bar{\mu}_{NH_4^+} \qquad \text{Equation 10}$$

Under these conditions Equations 5 and 8 can be equated and, neglecting activity coefficients:

$$-\Delta pH = \log([NH_4^+]_{in}/[NH_4^+]_{out}) \qquad \text{Equation 11}$$

Thus for the same reasons that amines at high concentrations specifically dissipate the ΔpH component of Δp, at low concentrations the ratio of amine inside/outside can be used to measure ΔpH. In the latter case conditions must be chosen so that the system is at steady-state where all net ion fluxes other than H^+ are zero. This is achieved by using low concentrations of amine and illuminating for several minutes to reach the steady-state condition. Methods based on this principle are given in Section 3.2.

3.1.5 *Generation of ΔpH and $\Delta\psi$ by Imposed Ion Gradients*

Transient ΔpH and $\Delta\psi$ can also be generated without the participation of photosynthetic reactions by imposing pH and ion gradients across the thylakoid. By pre-incubating thylakoids at pH 5 and then rapidly increasing the pH of the suspending medium to pH 8, a ΔpH of -3 units can be generated across the membrane. This is the well known acid/base jump and its use for inducing ATP synthesis in the absence of light is given in Section 4.3.

The generation of significant $\Delta\psi$ across the thylakoid by using photosynthetic processes is difficult to achieve in the laboratory without access to apparatus capable of generating short (μsec) flashes of light. However, quite large $\Delta\psi$ can be generated in the dark by imposing a diffusion potential across the thylakoid membrane. This can be done by suspending valinomycin-treated thylakoids in a medium low in K^+ and Cl^-, and then rapidly adding KCl (7). Under these conditions, the permeability of the thylakoid to K^+ is much greater than to Cl^-, and after KCl addition K^+ ions begin to diffuse into the thylakoid (down their concentration gradient) at a faster rate than Cl^- ions. This leads to a charge imbalance across the membrane, generating a $\Delta\psi$ of the same polarity (inside positive) as that created by photosynthetic processes.

If the membrane were completely impermeable to Cl^- ions, then K^+ entry would continue until the membrane potential was energetically equivalent to the initial K^+ concentration gradient across the thylakoid, that is, $\Delta\bar{\mu}_{K^+} = 0$, at which point the system would be at equilibrium. Under this assumption, the membrane potential developed is simply given by the Nernst relationship.

Neglecting activity coefficients, then from Equation 1,

$$\Delta\bar{\mu}_{K^+} = F\Delta\psi + 2.303RT\log([K^+]_{in}/[K^+]_{out}) = 0 \quad \text{Equation 12}$$

Therefore:

$$\Delta\psi = 2.303(RT/F)\log([K^+]_{out}/[K^+]_{in}) \quad \text{Equation 13}$$

At 25°C,

$$\Delta\psi = 59\log([K^+]_{out}/[K^+]_{in}) \quad \text{Equation 14}$$

Unfortunately, the thylakoid is not impermeable to Cl^- and the $\Delta\psi$ initially generated will be lower than that predicted by Equation 13. The discrepancy is not too serious when the concentration ratio of K^+ ions outside/inside is less than 10. However, for larger concentration ratios, the Goldman-Hodgkin-Katz flux equation should be used to estimate the initial $\Delta\psi$ developed by KCl addition (7):

$$\Delta\psi = \frac{RT\ln}{F}\left(\frac{\sum_C [C^+]_{out}P_C + \sum_A [A^-]_{in}P_A}{\sum_C [C^+]_{in}P_C + \sum_A [A^-]_{out}P_A}\right) \quad \text{Equation 15}$$

where P_C is the permeability coefficient of the cation C for the thylakoid membrane, and P_A the corresponding coefficient for the anion A. The summations are over all the cations and anions which are involved.

Again activity coefficients for all ions have been neglected in Equation 15. This equation also assumes that only univalent ions are out of equilibrium with respect to their concentrations and that all ions diffuse passively across the membrane. The problem with using Equation 15 is that the relative permeability coefficients P_C and P_A must be known and these vary depending on the state of the membrane and concentration of valinomycin etc. If only KCl is considered and the temperature is 25°C, then Equation 15 simplifies to:

$$\Delta\psi = 59\log\left(\frac{[K^+]_{out} + M[Cl^-]_{in}}{[K^+]_{in} + M[Cl^-]_{out}}\right) \quad \text{Equation 16}$$

where $M = P_{Cl}/P_K$, the ratio of permeabilities to Cl^- and K^+. Values for M can be derived from the literature (7,10). *Table 3* shows an example of the use of Equations 14 and 16 to estimate the diffusion potential set up on adding a KCl pulse to thylakoids.

Like the acid/base procedure for generating ΔpH, diffusion potentials are transient, and begin to decay immediately after salt addition. The half-time of decay depends on the general permeability properties of the membrane but is typically 5 sec (7). The use of diffusion potentials to enhance acid/base ATP synthesis is illustrated in Section 4.3.

Table 3. Diffusion Potentials Developed on Adding KCl to Thylakoids.

KCl concentration (mM)		*Initial $\Delta\psi$ developed (mV)*	
Before salt addition	*After salt addition*	*Nernst*	*Goldman*
1	3	28	26
1	10	59	52
1	100	118	82
1	1000	177	89

For each KCl concentration, a value is given for the Nernst (Equation 14) and Goldman (Equation 16) relationship. For the latter, a value of M = 0.03 was assumed (6).

3.2 9-Aminoacridine Fluorescence Technique for Measuring ΔpH

This method, developed by Schuldiner *et al.* (11), is a simple optical technique for determining ΔpH. The method relies on the principle outlined in Section 3.1.4 that the protonated form of amine will distribute itself across the thylakoid membrane in response to ΔpH (Equation 11). This distribution can be measured by assuming that 9-aminoacridine behaves as a simple amine, and that the dye in the intrathylakoid space is non-fluorescent whilst that in the outer aqueous phase is fluorescent. However, since both the protonated and unprotonated forms of the amine are fluorescent, a relationship between the total amine distribution and ΔpH across the thylakoid is required. This is derived as follows:

$$[A_t]_{in} = [A]_{in} + [AH^+]_{in} \qquad \text{Equation 17}$$

$$K_a = [A]_{in}[H^+]_{in}/[AH^+]_{in} \qquad \text{Equation 18}$$

where $[A]_{in}$ and $[AH^+]_{in}$ represent the concentrations of unprotonated and protonated amine in the intrathylakoid space and K_a is the association constant for the protonation reaction. Similar equations can be written for the outer aqueous phase. From Equations 17 and 18:

$$[AH^+]_{in} = [A_t]_{in}/(1 + K_a/[H^+]_{in}) \qquad \text{Equation 19}$$

By writing a similar equation for $[AH^+]_{out}$ and substituting in Equation 11, we obtain:

$$-\Delta pH = \log\left(\frac{[A_t]_{in}(1 + K_a/[H^+]_{out})}{[A_t]_{out}(1 + K_a/[H^+]_{in})}\right) \qquad \text{Equation 20}$$

Inspection of Equation 20 reveals that if $K_a \ll [H^+]_{in}$ and $K_a \ll [H^+]_{out}$, then Equation 20 is approximately given by Equation 21:

$$-\Delta pH = \log([A_t]_{in}/[A_t]_{out}) \qquad \text{Equation 21}$$

In other words, when the pK of the *mono*amine is much higher than the pH of the suspending medium, most of the amine exists in the protonated form and the distribution of total amine across the thylakoid will be close to the distribution of the protonated form. This is valid for 9-aminoacridine (pK_a = 9.9) providing the pH of the suspending medium is not much above pH 8.5.

To measure ΔpH without inducing uncoupling, the initial concentration of 9-aminoacridine in the medium is typically 0.5 – 20 μM. Over this concentration range, the observed intensity of fluorescence is proportional to the concentration of 9-aminoacridine in the medium. The fluorescence of 9-aminoacridine taken up can be regarded as zero, due to the very small cross-sectional area for absorbance in the thylakoid interior, and to concentration quenching effects. The uptake of 9-aminoacridine into the thylakoid is associated therefore with loss of fluorescence of the amine taken up. If we define F_i as the fluorescence in the absence of ΔpH and F_q as the fluorescence seen in the presence of ΔpH, the observed fluorescence quenching, Q, on generating ΔpH across the thylakoid is given by:

$$Q = (F_i - F_q)/F_i \qquad \text{Equation 22}$$

and:

$$[A_t]_{out} \propto 1-Q \; ; \; [A_t]_{in} \propto QV/v \qquad \text{Equation 23}$$

where v is the internal volume of the thylakoids and V is the total volume of the suspending medium. Thus, from Equations 21 and 23 we obtain:

$$-\Delta pH = \log[Q/(1 - Q)] + \log(V/v) \qquad \text{Equation 24}$$

The minus sign in Equation 24 arises because of the consistent sign convention adopted in Section 3.1.1. Experiments will yield a positive value for the right-hand expression of Equation 24 and thus ΔpH can be calculated providing that the ratio of intrathylakoid volume to external medium volume is known. In practice, this is generally not the case and values are taken from the literature.

3.2.1 *Measurement of ΔpH Generated by Coupled ATP Hydrolysis in the Dark*

(i) *Applications.* In terms of the equipment required, this is the simplest method for demonstrating the ability of thylakoids to generate a ΔpH. The method can be used to check that preparations are well coupled or to demonstrate that coupled ATP hydrolysis is protonmotive. The general advantages and disadvantages of this method are discussed in Section 3.2.2.

(ii) *Equipment.* Any commercial fluorimeter equipped with a monochromator on the excitation (measuring) beam is suitable. The wavelength for excitation should be 360 nm using as narrow a slit width as possible. If the machine is equipped with an emission monochromator, this is set to 490 nm. If there is no emission monochromator, then the photomultiplier must be blocked against excitation light (and chlorophyll fluorescence induced by the measuring beam) by using the filter combination given in Section 3.2.2.

(iii) *Method.* Thylakoids must be pre-illuminated in the presence of DTT to bring them into the thiol-reduced state and capable of catalysing ATP hydrolysis in the dark. The procedure is given in Section 2.2. After this modulation, keep the chloroplasts on ice for no longer than 30 min before use.

(a) Take 200 μl of pre-treated chloroplasts ($\sim 40-50$ μg chlorophyll) and add to 2.7 ml of assay medium AM1 (as shown in *Table 4*) in a 4 ml fluorescence cuvette.
(b) Mix well and place the cuvette in a strong light ($50-200$ W m^{-2}) for 15 sec.

Table 4. Media for the Assay of 9-Aminoacridine Fluorescence Quenching Induced by ATP Hydrolysis.

AMI	*ATP solution*	*Others*
0.33 M sorbitol	10 mM ATP	1 mM DCMU (in ethanol)
5 mM $MgCl_2$	pH 8.0	
20 mM KCl		1 M NH_4Cl
20 μM 9-aminoacridine		*or*
30 mM Tricine-KOH		100 μg nigericin ml^{-1} (in ethanol)
pH 8.0		

Nigericin is obtainable from Sigma Co.

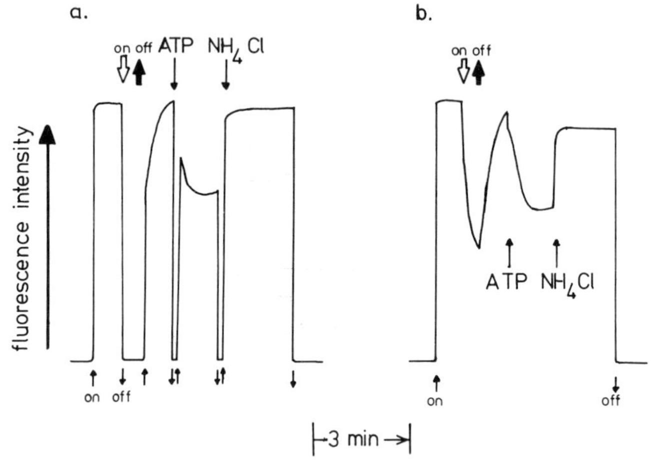

Figure 6. Quenching of 9-aminoacridine fluorescence induced by coupled ATP hydrolysis. The experiment was performed on either **(a)** an unmodified Perkin-Elmer 3000 spectrofluorimeter, or **(b)** a laboratory constructed fluorimeter [Section 3.2.2(ii)]. ATP and NH_4Cl were added to final concentrations of 0.33 mM and 10 mM, respectively. Lower arrows represent switching the measuring beam on or off; upper thick arrows represent switching the actinic beam on or off.

The light source for the thiol-modulation procedure may be used for this. The second illumination is to reactivate CF_0-CF_1 (see Section 4.1).

(c) Add 10 μl of 1 mM 3-(3,4-dichlorophenyl)-1,1-dimethylurea (DCMU) to the cuvette, mix well and place in the fluorimeter. The fluorescence signal should rise and stabilise within 60 sec.

(d) Add 100 μl of 10 mM ATP to the cuvette, mix and observe the fluorescence. As shown in *Figure 6a*, the signal will decrease as ΔpH develops in response to hydrolysis of the ATP.

(e) When the signal has stabilised, remove the cuvette from the fluorimeter, add 30 μl of 1 M NH_4Cl, mix well and replace the cuvette. The observed fluorescence should return to a level slightly below that of the initial starting value. The difference is caused by chemical quenching of 9-aminoacridine fluorescence by ATP.

The experiment is best performed using an apparatus with a stirred cuvette to which additions can be made during the measurement of fluorescence (see Section 3.2.2 and *Figure 6b*). In this case, the chemical quenching effects of ATP can often be seen as a rapid decrease of fluorescence which is not reversed by addition of the uncoupler. The ΔpH-dependent quenching appears as a slower second phase which is reversed by uncouplers (*Figure 6b*).

(iv) *Calculations.* ΔpH can be calculated directly from Equations 22 and 24. The value for F_i should be the *final* level of fluorescence seen in the presence of ATP plus uncoupler to allow for the chemical quenching effects of ATP. The value for F_q is the steady-state fluorescence level in the presence of ATP but without uncoupler present. The internal chloroplast volume (v) can be taken as 10 μl (mg chl)$^{-1}$ (12) unless measured (see Section 3.3) whilst the value for V is 3 ml.

(v) *Troubleshooting.* The failure to observe uncoupler-reversible 9-aminoacridine fluorescence quenching is nearly always due to poor chloroplast material. Check that the preparation is capable of uncoupler-stimulated ATP hydrolysis by the methods of Section 4. If the initial fluorescence level is below that of the final level in the presence of uncoupler, then the measuring beam may be causing generation of ΔpH *via* electron transport. This should not occur if the method given here is adhered to since the addition of DCMU will prevent methyl-viologen-dependent non-cyclic electron flow. If for any reason DCMU cannot be added, or cofactors of cyclic electron flow have been used during the thiol-modulation procedure, then this problem can be overcome by reducing the intensity of the measuring beam with Wratten 96 neutral density filters.

3.2.2 *Measurement of ΔpH Generated by Light-dependent Electron Transport*

(i) *Applications and advantages.* This is the commonest method for measuring steady-state ΔpH generated by electron transport during continuous illumination (11,13–16). Once the fluorimeter is set up, this method is simple and reproducible; experiments are also inexpensive to run. The technique is widely used, allowing a ready comparison of results with those in the literature. The method gives a real-time assay of ΔpH and, because quenching is logarithmically related to ΔpH, is sensitive in the range ΔpH = −2 to −3.5.

There are several disadvantages, most of which are concerned with the quantitative intepretation of the results. The internal volume of the thylakoids is rarely known under the conditions of the assay, and values generally have to be assumed. This can lead to systematic errors in the estimation of ΔpH which can be compounded if the thylakoid volume changes during the experiment. However, it has been found that thylakoid volume is generally rather constant and the errors associated with volume changes can generally be ignored (12).

Doubts have been raised as to whether 9-aminoacridine distributes across the thylakoid in a manner expected from theory (15). Binding, aggregation and micelle formation have been suggested to occur within the thylakoid, all of which will tend to cause ΔpH to be overestimated. Aggregation of 9-aminoacridine in the thylakoid is diminished if low concentrations (<1 μM) are used (14).

(ii) *Equipment.* The basic requirement is a fluorimeter equipped with two light sources. One (the measuring beam) is used conventionally at low intensities to excite

Table 5. Optical Filter Combinations for the Measurement of 9-Aminoacridine Fluorescence.

Measuring beam	*Actinic beam*	*Photomultiplier*
Balzers Calflex C	Balzers Calflex C	Corning 4-96 (4 mm)
Corning 7-39 (4 mm)	Corning 2-62 (2 mm)	Wratten 45
	or	
	Wratten 29	

Wratten filters are obtained from Kodak Co. Ltd. (UK address Manulife House, Marlborough St., Bristol, BS1 3NL); Corning filters from Corning Glassworks, Corning, NY 14830, USA (UK distributor: Precision Optical Instruments, Stratford Rd., Solihull, W. Midlands); Balzers filters from Balzers Aktengesellschaft (UK address: Balzers High Vacuum, Northbridge Rd., Berkhamsted, Herts HP4 1EN). The Calflex C heat filter should be placed nearest the light source.

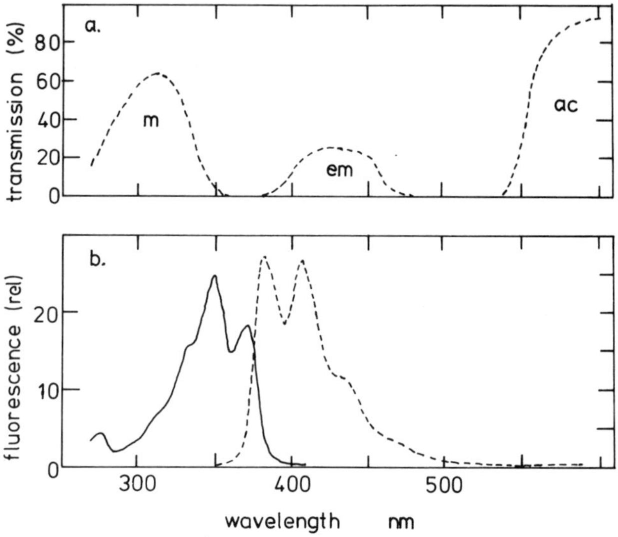

Figure 7. **(a)** Transmission spectra of the filter combinations given in *Table 5*: m, measuring beam (Corning 7-39); em, emission (Corning 4-96 plus Wratten 45); ac, actinic (Corning 2-62). **(b)** excitation (solid line) and emission (dotted line) spectra of 20 μM 9-aminoacridine measured on a Perkin-Elmer 3000 spectrofluorimeter at 2.5 nm slit width.

9-aminoacridine fluorescence. This light beam should be too weak to induce any significant ΔpH and should have a stabilised power supply to minimise fluctuations in the resulting fluorescence signal. The second light source is used at high intensities (>100 W m^{-2}) to drive photosynthetic reactions and need not have a stabilised supply. The optics must be arranged so that the photomultiplier only detects the fluorescence due to the measuring beam, and does not detect the actinic beam, either directly or through any fluorescence excited by the actinic light. This can be arranged by chopping the measuring beam and employing a phase-sensitive amplifier after the photomultiplier (Chapter 4). However, a simpler and cheaper method is to use complementary optical filters. The actinic light is filtered through filters that transmit only above 600 nm and the photomultiplier is blocked by filters that have no transmission in this region. The 9-aminoacridine fluorescence to be passed to the photomultiplier can be separated from the measuring beam by employing complementary filters in the excitation and emission pathways (see *Table 5* and *Figure 7*).

Perhaps the easiest way to obtain the apparatus is to adapt a commercial fluorimeter to accept an actinic light source. A hole can be cut in the cuvette housing of many fluorimeters, allowing an externally located 250 W tungsten-halogen light (e.g. from a slide projector) to be focussed onto the cuvette. Place a 50 × 50 mm Corning 2-62 red-transmitting filter over the hole and seal the edges with black insulation tape. Alternatively, cut a circular hole in the cuvette housing and introduce the actinic light using a 10 – 13 mm flexible fibre-optic light guide. Again seal around the edges of the cut hole to exclude stray light. A 10 or 25 mm diameter Corning 2-62 filter can then be taped directly over the cuvette end of the fibre-optic and the light guide secured so as to illuminate the cuvette at right angles to the emission pathway. The photomultiplier should be protected from the actinic light by placing a Corning 4-96 filter in the emission pathway.

There may be objections from other users to drilling holes in shared equipment and in this case, a simple yield fluorimeter may be constructed as discussed in Chapter 4 and Section 3.2.2 (vi) below. It is an advantage to build in a magnetic stirrer below the cuvette holder. If the budget is tight, it is not necessary to use a red-sensitive photomultiplier tube for the detection of 9-aminoacridine fluorescence. Any general purpose tube such as Thorn-EMI 6097B is suitable at a fraction of the cost. We have successfully used 6097B tubes and tube housings salvaged from disused scintillation counters. The dynode chain of the tube needs to be rewired to provide negative-polarity d.c. current instead of positive-polarity a.c. current. A simple resistive network is suitable (see Chapter 4, and 'Photomultipliers', a publication available from Thorn-EMI, Ruislip, Middlesex, UK). The photomultiplier power supplies in most scintillation counters are of positive-polarity and cannot be re-used.

The filter combinations for 9-aminoacridine fluorescence detection in a simple yield fluorimeter are given in *Table 5*, and *Figure 7* shows their transmission characteristics together with the excitation and emission spectra of 9-aminoacridine. Kodak Wratten filters are purchased as gelatin sheets that can be cut to size and mounted in 50 × 50 mm slide holders. If fibre-optics are used, mount the Wratten filters between circular glass cover slips. Glass filters can be purchased in 50 × 50 mm form or, for fibre-optics, in 10 or 25 mm circular form.

When setting up the apparatus, check that switching on the actinic light does not cause a response from the photomultiplier when an empty cuvette is in the fluorimeter. If there is a response, the most likely reason is a leak of white light to the photomultiplier tube. The golden rule to obey when setting up is that the only light to enter or leave the system must be through the filters. Pay particular attention to leaks around the edges of the filter holders or through cracks in the apparatus. If in doubt, seal it with black insulation tape. The other point to note is that the measuring beam must be of low intensity so that it does not drive photosynthesis. The intensity should be about 0.25 W m^{-2} but is best set using chloroplasts under experimental conditions.

(iii) *Method*.

(a) Place 2.5 ml of assay medium AM2 (*Table 6*) into a 4 ml fluorescence cuvette and add intact chloroplasts or thylakoids containing 50 μg chlorophyll.

(b) Mix well and, after 30 sec (during which lysis of intact organelles occurs), add 0.5 ml of 2 M sorbitol and mix well.

Table 6. Media for the Assay of 9-Aminoacridine Fluorescence Quenching Induced by Coupled, Non-cyclic Electron Transport.

AM2		*Uncouplers*
6 mM $MgCl_2$	2 M sorbitol	1 M NH_4Cl
24 mM KCl		
20 μM 9-aminoacridine		*or*
0.12 mM methyl viologen		
600 units catalase ml^{-1} (Sigma type C40)		100 μg nigericin ml^{-1} (in ethanol)
30 mM Tricine-KOH		
pH 8.0		

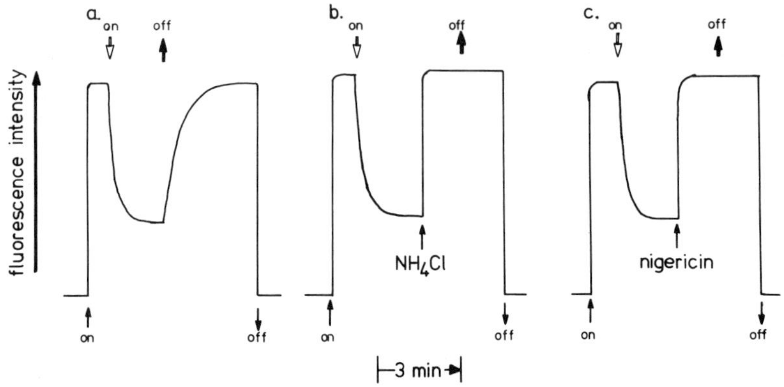

Figure 8. Quenching of 9-aminoacridine fluorescence by light-driven coupled electron transport. Reversal of quenching was induced by **(a)** switching off the actinic beam; **(b)** adding 10 mM NH_4Cl, or **(c)** adding 1 μg ml^{-1} of nigericin (final concentrations). See *Figure 6* and text for other details.

If intact organelles are required for the experiment, mix the AM2 and sorbitol *before* adding the chloroplasts. If the thylakoids have been thiol-modulated as described in Section 2.2, add 0.2 ml of the pre-treated suspension to 2.4 ml of AM2 and then add 0.4 ml of 2 M sorbitol. The pH and buffer of AM2 may be varied as required.

(c) Place the cuvette in the fluorimeter and observe the fluorescence in the presence of the measuring beam (set the photomultiplier voltage to 750 V).

On switching on the beam, the signal should rise rapidly then stabilise as shown in *Figure 8*. If the signal begins to decrease steadily, add 30 μl of 1 M NH_4Cl to uncouple the thylakoids. The fluorescence should then return to the starting value and remain stable. If it does, then the slow decrease was due to ΔpH generated by electron transport in the presence of the measuring beam. The intensity of the latter must be reduced by decreasing the voltage to the lamp or adding Wratten 96 neutral density filters until quenching is no longer observed in the presence of the measuring beam alone. Use a fresh sample of chloroplasts to set the intensity of the measuring beam and, once set, do not thereafter alter it.

(d) Take a fresh sample of chloroplasts, place in the fluorimeter and switch on the measuring beam.

(e) After 60 sec, switch on the actinic light and observe the fluorescence quenching (*Figure 8*). After 2 – 3 min, the signal should stabilise as ΔpH reaches its steady-state value. At this point, switch off the actinic light, or make additions of uncouplers such as NH_4Cl or nigericin (but not valinomycin) as shown in *Figure 8*. The signal should return to the starting level irrespective of the means by which ΔpH is dissipated.

(iv) *Calculations.* Calculate ΔpH from Equations 22 and 24 assuming an internal volume of 10 μl (mg chl)$^{-1}$ (12). If chloroplasts are intact during the assay, then ΔpH calculated will be the difference in pH between the intrathylakoid space and the suspending medium, not the stromal space. The true ΔpH will generally be higher than that calculated because the pH of the stroma is higher than that of the suspending medium (13).

(v) *Troubleshooting.* Under the given conditions, the signal measured in the absence of 9-aminoacridine should be less than 5% that of the signal in the presence of the fluorophore. If it is higher than this, check for light leaks in the system. The residual background signal is mainly due to imperfect filtering of the excitation beam.

Excessive noise in the chart recorder signal can arise for a number of reasons, but first check that the measuring beam light source has a properly stabilised power supply. Cheap sources such as slide projectors do not have stabilised power supplies to the lamp. However, if the lamp is a 12 V d.c. current type, this can be achieved simply by isolating the cooling fan from the lamp supply, and wiring the lamp to a 12 V automobile battery. Noisy signals may also arise if the measuring beam is too weak. A balance needs to be struck between reducing its intensity so as not to drive photosynthesis, yet keeping the intensity high enough so as not to require excessive amplification of the photomultiplier anode current. Noisy signals should not be a problem if a high-quality chart recorder capable of measuring 1 mV full-scale deflection is used.

(vi) *Upgrading.* The filters given in *Table 5* are inexpensive and easy to obtain and are suitable when the 9-aminoacridine concentration is in the range 5–20 μM. For lower levels of dye, the background signal due to imperfect filtering of the measuring beam from the fluorescence signal becomes excessive. Some improvement can be made by using narrow band-pass interference filters such as those recommended in (16). To measure ΔpH with dye concentrations below 1 μM, where aggregation effects are diminished (14), it is probably best to adapt a commercial fluorimeter and utilise its monochromators and high-quality amplifier system to detect fluorescence.

An excellent alternative to constructing a box-type fluorimeter (described in Chapter 4) is to base the apparatus around a well-type oxygen electrode (16). This provides a stirred, thermostatted, reaction vessel allowing simultaneous measurement of the rate of electron transport and ΔpH. Photophosphorylation can also be measured under identical conditions (see Section 4). Hansatech Ltd. manufacture a well-type electrode where the water jacket is made from aluminium and is light-tight. Ports in the jacket allow measuring and actinic light beams to be delivered *via* fibre-optic light guides (see *Figures 9* and *10*). A third port can be used to deliver fluorescence to the photomultiplier either directly (*Figure 10*) or *via* fibre-optics. The same filter combinations given in *Table 5* can be used in 25 mm circular form. The filter holders supplied by the manufacturer can be used to interface the fibre-optics with the reaction chamber. The data shown in *Figure 8* were obtained using the apparatus pictured in *Figure 10*. The apparatus

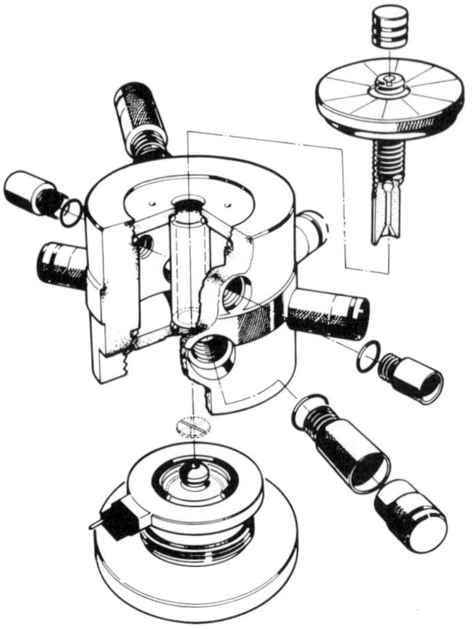

Figure 9. Exploded view of a Hansatech DW2 oxygen electrode (drawing courtesy of Hansatech Ltd.).

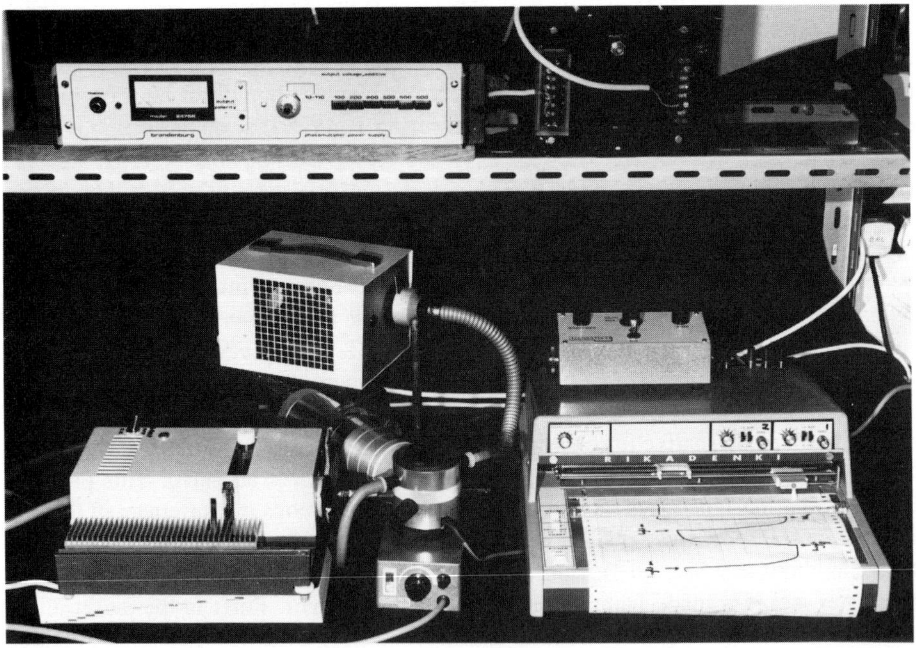

Figure 10. Photograph of a Hansatech DW2 electrode adapted to measure 9-aminoacridine fluorescence. The fibre optic delivers stabilised light (measuring beam) whilst the projector delivers unstabilised actinic light. The photomultiplier is interfaced directly with the circular filter holders that plug into ports in the electrode housing.

is a little more expensive to set up than a cuvette-based fluorimeter, but the cost is far outweighed by the resulting advantages of having a stirred temperature-controlled reaction vessel which can be used to measure the rate of electron transport and ATP synthesis under identical conditions to those used to determine ΔpH.

3.3 Radiolabelled Amine Technique for Measuring ΔpH

This method also relies on the net uptake of amines to measure ΔpH as described by Equation 21. In this case, the amine is labelled with ^{14}C or ^{3}H and uptake is measured directly after rapid separation of the thylakoids from the suspending medium by centrifugation through a layer of silicone oil. The uptake of amine is corrected for the suspension medium carried through the oil by the inclusion of an impermeable space marker such as sorbitol which is labelled with a second radioisotope. This double-label technique was developed by McCarty and associates (12,14) and has not been used by this author. The method given here is a simplified version of published techniques suggested by Professor R.E.McCarty (personal communication) which omits the second spacer marker.

(i) *Applications.* This simplified method is suitable for quantitative measurements of ΔpH values larger than −2 pH units under continuous illumination. Smaller values of ΔpH can be measured using the full double-label technique (12,14). The main advantage of this method is that it appears to be a more quantitative measure of ΔpH than the 9-aminoacridine technique, giving generally lower estimations that are independent of amine and chlorophyll concentrations (after volume changes are accounted for). Thus it appears that this method does not suffer from amine binding and aggregation effects that tend to cause the 9-aminoacridine technique to overestimate ΔpH. A second advantage is that the intrathylakoid volume can be estimated under identical conditions (12,14).

The main disadvantages are that it is a more time-consuming method, does not give a real-time measurement, and rates of electron transport and ATP synthesis are more difficult to measure under similar conditions.

(ii) *Equipment.* A microcentrifuge capable of rapid acceleration to full speed is required. In addition, the rotor must be illuminated during operation. The model recommended by McCarty is a Coleman TM 14K30, but a Beckman Microfuge 152 has also been used. Other modern microfuges are probably suitable, but the lid needs to be replaced with plexiglass or clear perspex or the lid interlock mechanism disabled to allow illumination from the top during operation. Any illumination system that gives an intensity of greater than 100 W m^{-2} at the rotor surface may be used [e.g. a 650 W tungsten/halogen floodlamp (14)]. Care should be taken that the centrifuge tubes are not heated during the illumination. The light should be filtered through a tank of water (or more preferably 0.5% $CuSO_4$ solution) and the lamp and centrifuge head cooled by fans.

(iii) *Method.*

(a) Place 0.1 ml of quench medium (*Table 7*) into 0.4 ml capacity polyethylene microcentrifuge tubes.

Table 7. Media for Silicone-oil Centrifugation Techniques.

Quench medium	*Silicone oil*	*AM3*
8.5% glycerol 2% trichloroacetic acid	2.2 parts (by weight) F(50) to 1 part SF96(50) *or* 9 parts AR 20 to 3 parts AR200	5 mM $MgCl_2$ 50 mM NaCl 50 μM PMS 25 μM [^{14}C]methylamine (10 μCi mol^{-1}) pH 8.2

Reagents are obtained as follows: F(50) and SF96(50) oils from Silicon Products Division, GEC, Waterford, New York, USA; AR20 and AR200 oils from Wacker Chemie, Munchen, FRG; [^{14}C]methylamine from Amersham International. For denser assay media, use a ternary silicone oil mixture containing F(50): SF96(50): Dow-Corning 702 diffusion pump oil in a ratio 0.9:1.0:0.6 by weight. PMS, phenazine methosulphate.

(b) On top of this, layer 0.1 ml of the silicone oil mixture (*Table 7*).

(c) Briefly centrifuge the tubes to remove air bubbles, keeping the temperature below 23°C to prevent possible inversion of the layers.

(d) In a separate tube, dilute an aliquot of chloroplasts with assay medium AM3 to give a final concentration of 0.1 mg chlorophyll ml^{-1}.

AM3 contains the radioactive amine. This can be ^{14}C or ^{3}H-labelled hexylamine or methylamine. Labelled hexylamine is now difficult to obtain commercially but can be used over a wide range of medium pH. Labelled methylamine should only be used at pH greater than 8.

(e) Take 0.1 ml portions of the diluted thylakoids (intact organelles will have been lysed) and carefully layer on top of the silicone oil.

(f) Place the tubes in the centrifuge and spin at the lowest possible speed (this is to even out the subsequent illumination).

(g) Switch on the light and illuminate for at least 30 sec. After this time, rapidly accelerate the microfuge to full speed and continue the illumination for a further 15 sec.

Thylakoids will pass through the silicone oil layer and be found in the lower aqueous phase. Consequently, the amine taken up will also be in the lower layer whilst the amine not taken up will be in the upper aqueous layer.

(h) Remove the tubes, discard the upper layer and most of the oil by aspiration, and cut the tubes just above the lower/oil aqueous interface.

(i) Place the bottoms of the tubes in screw-top scintillation vials and add 9 ml of toluene/Triton X-100 based scintillation fluid.

(j) Shake vigorously and count the samples.

(k) Carry out a control run in the dark to estimate the label carried over into the lower aqueous phase in the absence of ΔpH.

A correction for chloroplasts which remain in the upper layer can be made in parallel experiments in the absence of label. Take an aliquot of this upper layer for chlorophyll determination and prepare vials from the lower layer as before to estimate backgrounds and counting efficiencies. First count the vials for background then add 50 μl of radioactive AM3 (containing 1.25 nmol of labelled amine). Recount the vials and calculate the c.p.m. per nmol of amine.

(iv) *Calculations.* Calculate ΔpH using Equation 25:

$$-\Delta pH = \log(A_{t\ in}/A_{t\ out}) + \log(V/v) \qquad \text{Equation 25}$$

where $A_{t\ in}$ and $A_{t\ out}$ refer to amounts of amine in nmol, not concentrations.

Correct the c.p.m. for background and label carried over the dark. The latter correction may result in an underestimation of ΔpH if there is a stable ΔpH in the dark. However, for light-dependent ΔpH in excess of −2, the error is small. To estimate a ΔpH smaller than this, a more sophisticated method using a second label as a space marker should be used.

An internal thylakoid volume of 10 μl (mg chl)$^{-1}$ can be assumed. Alternatively, the volume can be measured directly by a similar double-label technique (12,14).

4. MEASUREMENT OF PHOTOPHOSPHORYLATION AND ITS PARTIAL REACTIONS

The measurement of CF_0-CF_1-catalysed ATP synthesis and hydrolysis is technically not difficult. However, recent research (17−20) has confirmed the long-held notion that CF_0-CF_1 is a highly regulated enzyme, capable of existing in several states of differing catalytic potential. Experiments therefore need to be designed with care to ensure that CF_0-CF_1 is in the correct state for the function being measured. Before describing the techniques available, the regulatory properties of the enzyme are summarised below.

4.1 **Principles**

4.1.1 *Thermodynamic Control of Photophosphorylation*

A central proposal of the chemiosmotic hypothesis is that ATP synthesis is catalysed by a reversible protonmotive ATPase (1). Experimental studies have confirmed the existence of such an enzyme in all energy-coupling membranes, including thylakoids. The function of the enzyme is to couple the processes of proton transport and ATP synthesis. The decrease in electrochemical potential experienced by protons passing from the intrathylakoid space into the outer medium is used to overcome the positive free energy change of ATP synthesis (ΔG_p). The mechanism by which this actually occurs is uncertain, but the pathway of H^+ efflux is probably *via* a channel through the membrane formed from the CF_0 subunits. CF_0-CF_1 is thus an energy transducer, interconverting $\Delta\bar{\mu}_{H^+}$ and ΔG_p. Since the system is reversible, net ATP hydrolysis is also protonmotive, translocating H^+ into the thylakoid against the $\Delta\bar{\mu}_{H^+}$. As in any reversible reaction, the system will tend to move towards equilibrium, when:

$$\Delta G_p - n\Delta\bar{\mu}_{H^+} = 0 \qquad \text{Equation 26}$$

where n is the number of protons translocated for each ATP molecule undergoing reaction. The value of n is not known for certain but is generally assumed to be 3. Under conditions typical for measuring *in vitro* ATP synthesis, ΔG_p has a value of around +36 kJ mol^{-1} and ATP synthesis will be in equilibrium when $n\Delta\bar{\mu}_{H^+}$ is therefore +36 kJ mol^{-1}, or $\Delta\mu_{H^+}$ = 12 kJ mol^{-1} assuming n = 3. (Note that $\Delta\mu_{H^+}$ is positive because of the sign convention adopted in Section 2.1 defining it in the direction of proton *uptake* into the thylakoid). A $\Delta\bar{\mu}_{H^+}$ of 12 kJ mol^{-1} is equivalent to Δp = 120 mV, or ΔpH = −2.

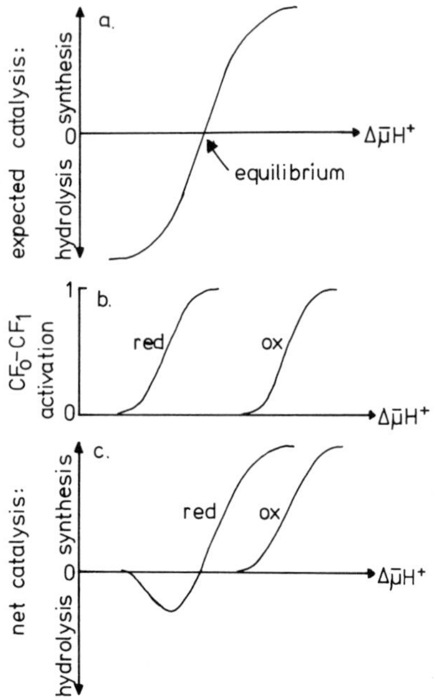

Figure 11. Predicted rate of ATP synthesis or hydrolysis as a function of $\Delta\bar{\mu}_{H^+}$, **(a)** if catalysis were controlled only by thermodynamic effects of $\Delta\bar{\mu}_{H^+}$, **(b)** if catalysis were controlled only by activation of CF_0-CF_1 by $\Delta\bar{\mu}_{H^+}$, **(c)** if both thermodynamic and kinetic activation effects controlled the rate of catalysis. ox represents CF_0-CF_1 in the thiol-oxidised form and red represents the enzyme in the thiol-reduced form. Note that the thiol-reduced form undergoes activation at lower values of $\Delta\bar{\mu}_{H^+}$ compared with the oxidised form.

As shown schematically in *Figure 11a*, the rate of ATP conversion (measured as net synthesis or hydrolysis) near to equilibrium should be controlled mainly by thermodynamic considerations. When $\Delta G_p < n\Delta\bar{\mu}_{H^+}$ then the system will be poised in the direction of net ATP synthesis whereas if $\Delta G_p > n\Delta\bar{\mu}_{H^+}$, then net ATP hydrolysis would be expected to occur.

There are theoretical reasons (21) for believing that the net rate of the reaction will be linearly dependent on $|\Delta G_p - n\Delta\bar{\mu}_{H^+}|$ in the region near equilibrium. In this region, the rate of ATP synthesis or hydrolysis should therefore be under thermodynamic control. However, as the difference between ΔG_p and $n\Delta\bar{\mu}_{H^+}$ increases, the rate cannot increase indefinitely because CF_0-CF_1 is an enzyme with a finite turnover capacity. There must therefore be regions far from equilibrium where the net rate is limited by the kinetic properties of the enzyme and independent of the thermodynamic properties of the system. The theoretical dependence of the rate of ATP synthesis or hydrolysis on $\Delta\bar{\mu}_{H^+}$ will therefore be an S-shaped curve as shown schematically in *Figure 11a*.

In practice, the rate of ATP synthesis is observed to increase up to the highest ΔpH that can be generated by coupled electron transport ($\sim\Delta pH = -3.5$ or $\Delta p = 200$ mV). Saturation of the rate is only observed under special conditions (acid/base jumps) when unusually high levels of ΔpH can be generated (17). ATP synthesis does not become independent of ΔpH until $\Delta pH < -4$ ($\Delta p > 240$ mV).

4.1.2 *Kinetic Control of Photophosphorylation*

The theoretical curve of *Figure 11a* assumes that the specific activity, or V_{max} of CF_0-CF_1 is independent of $\Delta\bar{\mu}_{H^+}$. Unfortunately, this is not the case for the chloroplast enzyme or for that from photosynthetic bacteria. Current research suggests that CF_0-CF_1 can exist in two (or more) states, one of which is catalytically inactive. The two states are interconvertible, but the equilibrium distribution of CF_0-CF_1 between these two states depends on the magnitude of $\Delta\bar{\mu}_{H^+}$ across the thylakoid membrane (17). When $\Delta\bar{\mu}_{H^+}$ is zero, the enzyme molecules are all in the inactive state and no catalytic activity is therefore observed. Upon increasing $\Delta\bar{\mu}_{H^+}$, the active state is progressively favoured as shown in *Figure 11b*. Thus as $\Delta\bar{\mu}_{H^+}$ increases, CF_0-CF_1 is progressively activated.

The situation is made more complex by the fact that the enzyme can also exist in two chemical forms depending on the redox state of thiol groups on the γ subunit of CF_1 (22). When these groups are oxidised, the $\Delta\bar{\mu}_{H^+}$ required to achieve significant activation of CF_0-CF_1 seems to be much higher than that required thermodynamically to poise the reaction in the direction of net ATP synthesis (19) (compare *Figure 11a* and *b*). Under these conditions, the rate of ATP synthesis is controlled kinetically by the activation state of the enzyme, and net ATP hydrolysis is not possible. However, when the γ thiol groups are reduced, then the $\Delta\bar{\mu}_{H^+}$ required to activate CF_0-CF_1 is less than that required to poise the reaction in the direction of ATP synthesis. Under these conditions, the rate of synthesis may be controlled by the thermodynamic properties of the system and when $\Delta\bar{\mu}_{H^+}$ is low, net ATP hydrolysis is possible.

The picture given above is undoubtedly an oversimplification of the complex regulatory properties of CF_0-CF_1 but broadly accounts for the experimental observations obtained in several laboratories which are summarised in *Figure 11c* and below.

(i) ATP synthesis appears to be preceded by activation of CF_0-CF_1 (17).

(ii) ATP hydrolysis is only observed when the enzyme is in the thiol-reduced state and requires the presence of a low $\Delta\bar{\mu}_{H^+}$ across the thylakoid (23).

(iii) Conversion of CF_0-CF_1 from the oxidised to the reduced form lowers the $\Delta\bar{\mu}_{H^+}$ required to observe a particular rate of ATP synthesis (20) [including the threshold ΔpH at which ATP synthesis is first observed in an acid/base transition (19)].

The above discussion refers to steady-state conditions of ATP synthesis or hydrolysis and $\Delta\bar{\mu}_{H^+}$. The situation is even more complex when sudden perturbations are made to the steady-state. For example, the rate at which CF_0-CF_1 reverts to an inactive state when $\Delta\bar{\mu}_{H^+}$ is suddenly collapsed seems to depend on which redox form the enzyme is in, being much slower when the enzyme is thiol-reduced. This leads to rather complex kinetics when uncouplers are added to chloroplasts catalysing ATP hydrolysis (23). Initially, the steady-state rate of hydrolysis is low, and uncoupling increases the rate as thermodynamic constraints are relaxed (see *Figure 13b*). However, the uncoupled rate is then observed slowly to decrease as CF_0-CF_1 reverts to the inactive state (the active state being unstable in the absence of $\Delta\bar{\mu}_{H^+}$). The practical consequences of all this are that the experimental conditions must be carefully selected for assay of CF_0-CF_1-dependent reactions, particularly ATP hydrolysis.

4.1.3 *Energy-transfer and Other Inhibitors of CF_0-CF_1*

Inhibition of ATP synthesis by uncouplers (Section 2.1.4) is caused by dissipation of $\Delta\bar{\mu}_{H^+}$ due to accelerated non-phosphorylating H^+ efflux from the thylakoid. Uncoupling is associated with an increase in the rate of electron transport. There are a large number of other inhibitors of ATP synthesis that interact more directly with the enzyme, and their effects include stimulation of $\Delta\bar{\mu}_{H^+}$ and depression of the rate of electron transport. These inhibitors are generally called energy-transfer inhibitors, and the more useful ones include dicyclohexyl carbodiimide (DCCD) (24), triphenyl tin (25) and tentoxin (26). The first two appear to act by blocking the H^+ channel in CF_0, thereby inhibiting both ATP synthesis and hydrolysis. The effects of tentoxin are complex, but at low concentrations ($< 10\ \mu M$), it seems to interact directly with the α and β subunits of CF_1. However, certain species of plants such as tobacco are naturally resistant to tentoxin. These energy-transfer inhibitors are useful as experimental tools to vary the activity of CF_0-CF_1 without affecting the coupling state of the membrane. In this respect, triphenyl tin and tentoxin are better inhibitors to use than DCCD, which has other non-specific inhibitory and uncoupling effects. The inhibition by triphenyl tin is reported to be reversed by treatment with reduced thiols (25).

4.2 **Measurement of Steady-state ATP Synthesis or Hydrolysis**

Four basic techniques are given all of which depend on direct or indirect assay of the reaction product. Conditions given are optimal, but ATP conversion can be measured at lower pH with reduced rates (20 – 30% at pH 7). Suitable alternative buffers are MES (pH 6 – 7), Hepes (pH 7 – 8) and glycylglycine (pH 8 – 9). Phenazine methosulphate or pyocyanin may generally be used as alternative cofactors of electron transport but ferricyanide causes oxidation of CF_1 (and DTT) and should not be used when ATP hydrolysis is to be measured.

4.2.1 *Inorganic Phosphate Assay for ATP Hydrolysis*

ATP hydrolysis is measured by assaying the inorganic phosphate released using a variation of the Fiske-Subbarow technique (3).

(i) *Applications and advantages.* This is an extremely simple method requiring no special apparatus and is suitable for teaching as well as research. A large number of assays can be run in a short time making the method useful for routine screening and establishment of inhibition curves. The assay is reliable for all but the lowest rates of CF_0-CF_1 catalysed ATP hydrolysis [$> 10\ \mu$mol ATP (mg chl)$^{-1}$ h^{-1}].

(ii) *Equipment.* An illumination source for the thiol-modulation procedure is required as given in Section 2.2. Any laboratory spectrophotometer will serve for colorimetric determination of phosphate at 800 nm. Make sure all glassware is clean to avoid contamination of samples by extraneous phosphate. Do not use phosphate-based detergents to clean the glassware.

(iii) *Method.*

(a) Add 0.9 ml aliquots of assay medium AM4 (*Table 8*) to 15 ml centrifuge tubes and place the tubes in a shaking water bath set at 25°C.

Table 8. Media for Assay of ATP Hydrolysis by Inorganic Phosphate Release.

AM4		*Pi assay*
2.5 mM $MgCl_2$ 2.5 mM ATP 1.25 mM NH_4Cl 30 mM Tricine-KOH pH 8.0	5% trichloroacetic acid	2.5% ammonium molybdate in 5N HCl 40% $SnCl_2$ in conc HCl (dilute 0.25 ml of this into 40 ml of H_2O just prior to use)

Prepare AM4 just prior to use.

(b) To measure ATP hydrolysis, thylakoids must first be brought into the activated, thiol-reduced state. Carry out the procedure given in Section 2.2 and, immediately after the 6 min pre-illumination, add an 0.1 ml aliquot to a centrifuge tube containing AM4. If several samples need to be taken, continue the pre-illumination taking 0.1 ml aliquots of chloroplasts every 15 sec.

If the thiol-reduced chloroplasts have been stored in the dark after the thiol-modulation procedure, they will remain in the reduced form but may have partially deactivated. In this case, re-illuminate the thylakoids for 30 sec at a light intensity of >50 W m^{-2} to reactivate CF_0-CF_1.

(c) Incubate the thylakoids in AM4 for exactly 4 min and then stop the reaction by adding 0.5 ml of 5% trichloroacetic acid (TCA). As a control, include a zero-time tube where the TCA was added before the chloroplasts.

(d) Use a bench centrifuge to pellet the precipitated membranes then take duplicate 0.5 ml aliquots of the supernatant and add to 15 ml test tubes containing 2.0 ml of distilled water.

(e) For the phosphate calibration curve, prepare a 10 mM solution of $NaH_2PO_4 \cdot 2H_2O$ (do not use the disodium salt because the water content of the crystals can be variable).

(f) Using a clean, precision glass syringe, add 0–20 μl (0–200 nmol phosphate) to a series of tubes containing 2.5 ml H_2O. (Better still, repeat the experiment but omit ATP from the assay medium then prepare the calibration tubes by adding the phosphate standards to 2 ml of H_2O and 0.5 ml of supernatant).

(g) Inorganic phosphate is determined as the reduced phosphomolybdate complex. Prepare stock solutions of 2.5% ammonium molybdate in 5 N HCl and 40% $SnCl_2$ in concentrated HCl. Other reductants can be purchased from chemical companies but $SnCl_2$ is inexpensive and both the molybdate and $SnCl_2$ stocks are stable at room temperature for up to 1 year. Just prior to the assay, dilute 0.25 ml of the concentrated $SnCl_2$ into 40 ml of H_2O and use within 1 h.

(h) Each test tube should now contain 2.5 ml of solution to be assayed for inorganic phosphate. Add 0.6 ml of molybdate and mix, add 0.25 ml of the diluted $SnCl_2$, mix and allow 15 min for the blue colour to develop.

(i) Read the samples at 800 nm in a spectrophotometer against a reagent blank containing 2.5 ml of water, molybdate and $SnCl_2$.

(iv) *Calculations.* Construct a calibration curve for the standards and use this to determine the nmoles of phosphate per test tube in the samples. Subtract from this the value found for the zero-time control which represents the phosphate contamination (mainly due to the phosphate content of the ATP, normally <2%). Use the corrected results to calculate ATPase activity in the assay as μmol phosphate released (mg chl)$^{-1}$ h^{-1}.

(v) *Troubleshooting*. Under the above conditions, typical rates of ATP hydrolysis are 120 – 250 μmol phosphate released (mg chl)$^{-1}$ h^{-1} giving an OD_{800} of 0.3 – 0.7. High rates but low OD_{800} indicates interference in the colorimetric assay. Sugars such as sorbitol (>50 mM) produce this effect and should not be included in the assay medium. Ferricyanide produces a brown precipitate in the assay but does not interfere at 800 nm (though ferricyanide should not be used for other reasons given above).

The observed rate of hydrolysis is close to the maximum uncoupled rate due to the presence of 1 mM NH_4Cl. This level of amine is sufficient partially to uncouple the thylakoids, accelerating the rate of hydrolysis but allowing a low ΔpH to be generated so that CF_0-CF_1 remains fully activated. The purpose of adding the uncoupler is to increase the sensitivity of the assay and to allow the full catalytic potential of the enzyme to be expressed. As shown in *Figure 12a*, high rates of hydrolysis in the presence of 1 mM NH_4Cl remain linear over the time of the assay (4 min). Higher concentrations of NH_4Cl do not further accelerate the rate but often cause the rate to become non-linear (see *Figure 13b*). This can lead to an apparent optimal concentration of NH_4Cl of around 1 mM when the assay time is 4 min, as shown in *Figure 12b*. However, care should be taken when the rate of hydrolysis is expected to be low. Under these conditions, a concentration of 1 mM NH_4Cl in the assay may be too high to allow the thylakoids to generate a ΔpH large enough to keep CF_0-CF_1 active over the time course of the assay.

4.2.2 *pH Electrode Technique for Measuring ATP Synthesis or Hydrolysis*

At pH 8 and excess Mg^{2+}, ATP synthesis or hydrolysis results in the scalar consumption or production of nearly 1 H^+ ion per ATP (27) as shown by the following reaction:

$$[Mg\cdot ADP]^- + PO_4^{2-} + H^+ = [Mg\cdot ATP\cdot H]^{2-}$$

When thylakoids are suspended in lightly buffered media, these scalar [H^+] changes can be detected as small changes in pH and used to calculate the rate of the reaction (23,27).

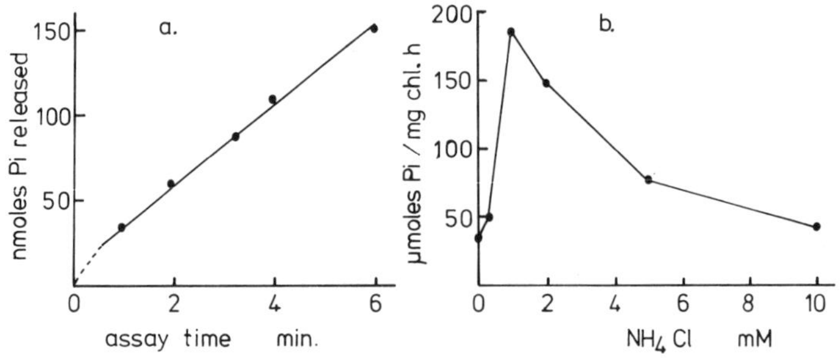

Figure 12. ATP hydrolysis measured by the inorganic phosphate technique. **(a)** Amount of phosphate released as a function of the assay time. **(b)** Effect of varying the concentration of NH_4Cl in the assay on the observed rate over 4 min.

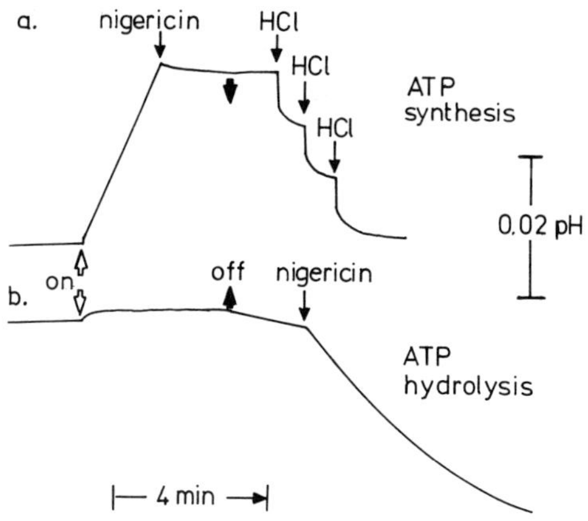

Figure 13. **(a)** ATP synthesis and **(b)** hydrolysis measured by the pH electrode technique. Nigericin was added to a final concentration of 1 μg ml^{-1}.

(i) *Applications and advantages.* This is the simplest method that provides a real-time measurement of ATP synthesis or hydrolysis. The technique is therefore ideally suited to rate determinations and kinetic studies, particularly for ATP hydrolysis. Little special apparatus is required and running costs are low.

(ii) *Apparatus.* A sensitive pH meter capable of measuring ± 0.002 pH units is required. It must have a chart recorder output and the output needs to be backed-off (i.e., have most of the d.c. component of the output removed) so that pH changes of 0.05 units can be displayed on the recorder. A pH meter that has an expandable display (2 pH units full-scale deflection) is suitable. The 'buffer control' adjustment on the pH meter may be used to back-off the recorder output; otherwise a separate back-off voltage will need to be applied to the recorder. The recorder needs to be sensitive, with a full-scale deflection of 1 – 10 mV.

Measurements can be made in a well-type oxygen electrode chamber equipped with an illumination source. A micro-combination pH electrode can be inserted through the top. For prolonged use, see (vi) below.

(iii) *Method.* The following should be done under dim room light.

(a) Take chloroplasts containing 75 μg chlorophyll and add to 2.5 ml of the lightly buffered assay medium AM5 (*Table 9*).

(b) Allow 30 sec for any intact organelles to become lysed, then add 0.5 ml of 2 M sorbitol.

(c) Insert the pH electrode and observe the pH on the meter, making sure that the suspension is well stirred. Set the pH to around 8 using freshly prepared 100 mM KOH and HCl.

(d) Set up the recorder to measure the pH, then using the 'buffer control' on the pH meter or an external voltage source, back-off the signal whilst increasing the sensitivity of the recorder until pH changes of 0.05 units can be convenient-

Table 9. Media for Assay of ATP Synthesis or Hydrolysis by the pH Electrode Technique.

AM5	*AM6*	
6 mM $MgCl_2$	6 mM $MgCl_2$	2 M sorbitol
25 mM KCl	25 mM KCl	
0.1 mM methyl viologen	0.1 mM methyl viologen	
1.0 mM ADP	1.0 mM ATP	
1.0 mM NaH_2PO_4	6 mM dithiothreitol	
2.5 μM diadenosine pentaphosphate	2.5 μM diadenosine pentaphosphate	
2 mM Tricine-KOH	2 mM Tricine-KOH	
pH 8.0	pH 8.0	

Prepare media just prior to use.

ly recorded on the chart. The signal will probably drift with time but this should soon stabilise to an acceptable level.

(e) After noting the rate of drift, switch on the actinic light source.

As shown in *Figure 13a*, a small but rapid alkalisation is seen on switching the light on and this partly is due to net H^+ uptake by the thylakoids. If ADP is omitted from AM5, the net H^+ uptake ceases after 30 sec. In the presence of ADP and Pi, net H^+ consumption continues at a linear rate after 30 sec due to scalar consumption of H^+ by ATP synthesis.

(f) Continue the illumination long enough to assess the rate of the pH change in the linear phase.

Figure 13a shows that addition of an uncoupler such as nigericin rapidly inhibits the alkalisation of the medium, indicating that net ATP synthesis has been abolished by the uncoupler.

(g) Switch off the light and, after the signal has stabilised, calibrate the observed pH changes by injecting 10 μl of freshly prepared 10 mM KOH or HCl using a precision glass syringe.

ATP hydrolysis can be measured in a similar way. Thylakoids should be thiol-modulated *in situ* as follows.

(a) Add chloroplasts to the oxygen electrode chamber as before but use assay medium AM6 in place of AM5 (*Table 9*).

(b) Set the system up to measure pH changes at pH 8 then illuminate for 4 min at an intensity no greater than 100 W m^{-2}.

Upon turning off the light, net hydrolysis of ATP will begin and this can be detected as continuous acidification of the medium (*Figure 13b*).

(c) Allow the rate to become linear (when net H^+ efflux from the thylakoids has ceased). This rate represents coupled ATP hydrolysis. The uncoupled rate can be estimated after adding nigericin as shown in *Figure 13b*. The high initial rate of uncoupled hydrolysis will not be maintained because CF_0-CF_1 will slowly become inactive in the absence of a ΔpH across the membrane.

At the end of the experiment, soak both the pH and oxygen electrodes in dilute detergent (0.2% Decon 90, Sigma Co. or similar phosphate-free detergent) to remove protein and membrane contamination.

(iv) *Calculations.* The number of scalar protons produced or consumed per ATP converted is a complex function of both the pH and Mg^{2+} concentration in the medium (27). At pH 8 and excess Mg^{2+}, the stoichiometry is nearly 1 but at lower pH it must be calculated from the published dissociation constants (27). Calculate the rate of H^+ production from the calibrations of HCl and KOH, allowing for the drift (estimated from the dark drift in the case of synthesis and from the drift just prior to switching off the light in the case of hydrolysis). At pH 8, the rate of H^+ consumption or production is directly equivalent to the rate of ATP synthesised or hydrolysed (mg chl)$^{-1}$ h^{-1}.

(v) *Troubleshooting.* The major problems likely to be encountered are noisy traces and excessive drift of the signal in the dark. Both of these can be minimised by keeping the pH electrode clean as described above. Even so, the porous plug of combination electrodes will tend to become clogged and short electrode lifetimes can be expected. Occasional soaking of the electrode in trypsin solution sometimes helps.

To ensure linearity of the calibrations, the total pH change during the assay must be small (<0.05 units). When changing the conditions of the assay, ensure that the total buffering of the system is adequate to prevent larger changes (this also helps to reduce drift).

Electrode responses to light in the absence of chloroplasts are generally due to heating artifacts. Make sure the light source is well filtered (Section 2.2) and the reaction chamber thermostatted.

(vi) *Upgrading.* The problems of electrode drift and noise are greatly reduced by using separate pH and reference electrodes and employing mechanical stirrers to achieve high stirring rates. A suitable reaction chamber will have to be built to incorporate these features and several designs are available in the literature (23,28).

4.2.3 [^{32}]*P Technique for Measuring ATP Synthesis and* ^{32}*Pi* ⇌ *ATP Exchange*

(i) *Applications and advantages.* Like all radioisotope tracer techniques, the rate of incorporation of ^{32}Pi into ATP measures the absolute rate of ATP synthesis regardless of the rate of the back reaction. For this reason, this method can be used to estimate net rates of ATP synthesis (17,20) by arranging for the back reaction (hydrolysis) to be insignificant or it can be used to estimate the rate of synthesis during net hydrolysis (Pi ⇌ ATP exchange). By using a coupled enzyme assay to trap incorporated labelled ATP, the rate of synthesis can be measured at low [ADP] making the method suitable for determination of K_m (ADP). Large numbers of samples can be run in a short time. The technique does not provide a real-time measure of ATP synthesis however, so several samples need to be taken to establish rates. Since the rate of synthesis is independent of ATP/ADP ratios over a wide range, this is not usually a problem. The cost of ^{32}Pi is low and scintillation cocktails are not necessary for counting, so experiments are inexpensive. However, ^{32}P is a high-energy β emitter so precautions in handling the isotope are necessary (use 2 cm thick Perspex shields when diluting stock solutions).

(ii) *Apparatus.* Use a well-type oxygen electrode as the reaction chamber (Section 2.2). To avoid contamination of the chamber, radioactive solutions can be placed in disposable

Table 10. Media for Assay of ATP synthesis and Pi ⇌ ATP Exchange by the Incorporation of ^{32}Pi into ATP.

AM7	*AM8*	*Extraction media*
6 mM $MgCl_2$	6 mM $MgCl_2$	1. Freshly prepared solution (50 ml) of:
25 mM KCl	25 mM KCl	1.5 g ammonium molybdate
0.1 mM methyl viologen	2 mM ATP	2.5 ml conc HCl
1.0 mM ADP	0.2 mM ADP	0.5 ml triethylamine
0.6 mM NaH_2PO_4 (containing 0.5 μCi ^{32}Pi ml^{-1})	0.6 mM NaH_2PO_4 (containing 0.5 μCi ^{32}Pi ml^{-1})	2. 0.5 mM NaH_2PO_4
2.5 μM diadenosine pentaphosphate	2.5 μM diadenosine pentaphosphate	
4 units hexokinase ml^{-1}	35 mM Tricine-KOH	
15 mM glucose	pH 8.0	
35 mM Tricine-KOH		
pH 8.0		

Prepare all media just prior to use.

plastic tubes which are then inserted in the chamber. Use 16 mm tubes for Rank electrodes and 10 mm tubes for Hansatech electrodes. Stirring is achieved using a 6 mm Teflon-coated magnetic bar (Bel-Art Products) in each tube. These bars are easily decontaminated by rinsing in water after the experiment.

(iii) *Method.*

(a) To measure ATP synthesis, add 0.9 ml of assay medium AM7 (*Table 10*) to 16 or 10 mm disposable test-tubes.

AM7 contains hexokinase and glucose to regenerate ADP and trap incorporated ^{32}P as labelled glucose-6-phosphate. This prevents any ATP hydrolysis. AM7 also contains diadenosine pentaphosphate to inhibit chloroplast adenylate kinase which together with the added hexokinase would slowly convert the added ADP to AMP.

(b) Add chloroplasts containing 20 μg of chlorophyll and allow at least 60 sec for equilibration.

The effects of uncouplers and energy transfer inhibitors can be studied by adding them at this stage. Some inhibitors such as tentoxin are slow to react with CF_0-CF_1 and longer equilibration times (5 min) must be given.

(c) After the dark equilibration period, illuminate for 60 sec. Light intensities of 250 W m^{-2} are required for saturation. Stop the reaction by adding 0.2 ml of 20% TCA. Go to step (f).

(d) To measure ^{32}Pi ⇌ ATP exchange during net hydrolysis, place 0.9 ml of assay medium AM8 (*Table 10*) into 15 ml disposable tubes and place in a shaking water bath at 25°C. Chloroplasts must be thiol-modulated as described in Section 2.2.

(e) 5 sec after pre-illumination, add 0.1 ml of the pre-treated chloroplasts to each tube and incubate for 4 min. Stop the reaction with 0.2 ml of 20% TCA.

(f) For both synthesis and exchange, several controls should be run where the TCA is added before the chloroplasts. To estimate incorporated ^{32}P, the unreacted ^{32}Pi must be removed by precipitation as the phosphomolybdate complex (29).

(g) To each tube, add 1.2 ml of freshly prepared molybdate/triethylamine (*Table 10*) and leave the tubes for 15 min for the precipitate to form fully.
(h) Spin the tubes in a bench centrifuge to pellet the precipitate (and membranes), then carefully take 1.5 ml of the supernatent and add to a clean disposable centrifuge tube containing 1.5 ml of 0.5 mM unlabelled NaH_2PO_4. A second white precipitate will spontaneously form and the tubes should again be left for 15 min before spinning.
(i) Finally, take 1 ml duplicate aliquots of the supernatant after the second spin and add to 4 ml H_2O in minivials for counting.

^{32}P can be counted *via* Cerenkov radiation without the need for scintillation cocktails. The counts appear at the lower end of the energy spectrum and an appropriate energy window needs to be determined beforehand (the tritium channel on most scintillation counters is suitable for this purpose).

The counts due to residual ^{32}Pi carried through the precipitation steps are estimated from the controls, which can also be used to determine counting efficiencies by adding an internal standard. Count the controls for carry-over, then add 0.1 ml of 10× diluted assay medium (which contains a known amount of ^{32}Pi) and recount.

(iv) *Calculations.* Calculate the counting efficiencies as c.p.m. per nmol ^{32}P by subtracting the carry-over counts of the control vials from the counts obtained after adding diluted assay medium. Correct the experimental sample vials for carry-over and convert the c.p.m. to nmol ^{32}P incorporated using the counting efficiencies determined above. Now calculate the rate of ^{32}Pi incorporation as μmol (mg chl)$^{-1}$ h^{-1} in the assay not forgetting the various dilutions during the precipitation steps.

(v) *Troubleshooting.* When using thiol-modulated chloroplasts, or if reductants have been added to the assay, the final solutions to be counted will appear blue due to reduction of residual phosphomolybdate in the vials. It is therefore essential that counting efficiencies be determined for each change of experimental conditions since colour quenching during counting may be changed.

The maximum rate of ATP synthesis is typically in the range 200–400 μmol (mg chl)$^{-1}$ h^{-1}, which will give 10–20% incorporation of the added ^{32}Pi in the assay. The observed activity in the vials will therefore be around 10 000–20 000 c.p.m. depending on the counting efficiency. Since the counts in the control vials (carry-over) are normally about 100, the rate of ATP synthesis can be accurately measured down to 10% of the maximum rate without changing the conditions. For lower rates or higher precision, either the assay time or chlorophyll concentration can be increased.

Sugars such as sorbitol interfere with the precipitation procedure and they should not be included in the assay medium.

Figure 16b shows some typical results obtained using a slightly modified procedure.

4.2.4 *Luciferin/Luciferase Technique for Measuring ATP and ADP*

This is the most sensitive of several enzymic methods for the direct assay of ATP and ADP. The assay depends on the following reactions:

$$\text{ATP} + \text{luciferin} + O_2 \rightarrow \text{AMP} + \text{PPi} + \text{oxyluciferin}^*$$

$$\text{oxyluciferin}^* \rightarrow \text{oxyluciferin} + \text{light}$$

Luciferase catalyses the oxidation of reduced luciferin in the presence of ATP, Mg^{2+} and oxygen. The initial product is oxyluciferin*, which is in an electronically excited state. Transition of the excited state to the ground state occurs with emission of a quantum of light at 560 nm (bioluminescence). By keeping the concentration of ATP well below the K_m for luciferase (50 μM), the rate of the overall reaction, and therefore the intensity of light emission, is proportional to the ATP concentration in the assay. Since low levels of light can be measured with a photomultiplier, high sensitivity is achieved and ATP concentrations as low as 10^{-11} M can be measured. Typically however, the [ATP] levels in the assay are normally in the range $10^{-8}-10^{-6}$ M.

(i) *Applications and advantages.* The major advantage of this method is its high sensitivity. Using a sampling technique, the ATP content of isolated intact chloroplasts can easily be measured under conditions typical for the measurement of CO_2 fixation. The ADP and AMP content of chloroplasts can also be measured after enzymic conversion to ATP using pyruvate kinase and adenylate kinase. The method can easily be adapted to measure *in vitro* ATP synthesis by thylakoids or ATP hydrolysis at low [ATP].

The chief disadvantage is the relatively high cost of the reagents and the need of a sensitive apparatus for measuring bioluminescence. The technique is generally suitable only for sampling methods and does not lend itself to real-time assays of ATP synthesis or hydrolysis.

(ii) *Equipment.* A light-tight cuvette housing equipped with a photomultiplier is required to measure light emission in the luciferase assay. Stirring and temperature control are not essential. Commercial luminometers such as the LKB-Wallac model 1250 are available, but a simple laboratory-constructed fluorimeter (Chapter 4) may be used. In this case, the measuring beam is not required and should be switched off. Remove any filters in front of the photomultiplier and adjust the monochromator (if present) to maximum slit width at 560 nm. Since the light levels to be detected are low, the apparatus must be absolutely light-tight during operation.

(iii) *Method.* To illustrate the technique, a method to analyse the ATP and ADP levels in isolated intact chloroplasts will be described.

(a) To 0.19 ml of resuspension medium RM1 (*Table 1*) in a centrifuge tube, add 10 μl of intact chloroplasts ($\sim$20 μg chlorophyll).

(b) Illuminate the tube for 1 min at a light intensity of greater than 50 W m^{-2} then add 0.2 ml of 10% TCA before terminating the illumination.

(c) Repeat the process with a second sample but maintain the chloroplasts in complete darkness throughout.

(d) To each tube, add 0.6 ml of Tris-acetate buffer (*Table 11*) and spin the tubes to remove precipitated membranes.

(e) Take 0.5 ml of the supernatant and neutralise it with 0.5 ml of 0.125 M KOH.

The neutralised solutions are asayed first for ATP, then for ADP. The luciferin/luciferase reagents are best obtained as a prepared kit such as LKB-Wallac Monitoring Reagent. Reconstitute the reagent according to the manufacturers instructions. Excess reagent can be split into small aliquots and frozen for future

Table 11. Media for the Assay of ATP and ADP by the Luciferin/Luciferase Technique.

Buffer	*ATP determination*	*ADP determination*
0.1 M Tris-acetate pH 7.75	0.1 μM ATP	0.1 M phosphoenol pyruvate
	LKB Monitoring Reagent	4 mg pyruvate kinase ml^{-1} (type III, Sigma) in Tris-acetate

Monitoring Reagent is obtained from LKB-Wallac Co., Turku, Finland. Store all media except buffer on ice.

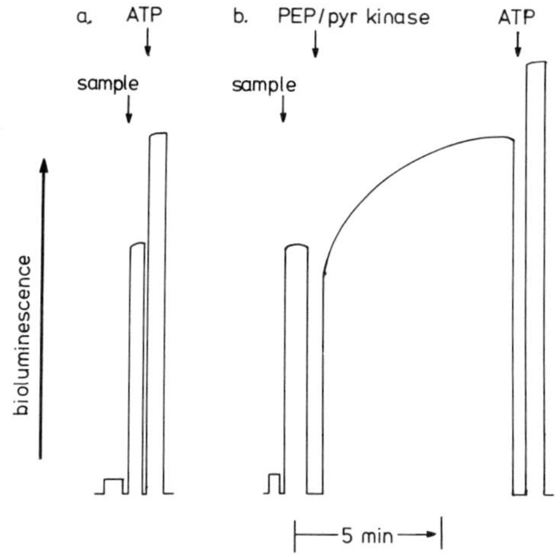

Figure 14. Determination of ATP and ADP by the luciferin/luciferase technique. See text for details. The ATP and ADP levels estimated in this experiment were 4.9 and 13.5 nmol (mg chl)$^{-1}$ in the dark and 10.7 and 10.6 nmol (mg chl)$^{-1}$ in the light.

use (do not repeatedly freeze and thaw the reagent or inactivation of the enzyme will result). Store the reagent to be used on ice.

(f) To 0.85 ml of Tris-acetate buffer, add 50 μl of Monitoring Reagent. (LKB-Wallac recommended the use of 200 μl, but for ATP concentrations in the range $10^{-8}-10^{-6}$ M, the lower level of 50 μl is quite adequate and results in considerable savings in the cost of the experiment).

(g) Observe the luminescence signal in a suitable luminometer (*Figure 14*). Add 100 μl of neutralised supernatant and observe the signal. The increased light emission is due to the ATP content of the supernatant. Calibrate this by adding 5 μl of 0.1 μM ATP using a precision glass syringe (*Figure 14a*).

(h) To analyse ADP, repeat the procedure with a fresh sample but in place of the calibration, add 10 μl of 0.1 M phosphoenol pyruvate and 40 μg of pyruvate kinase (*Table 11*).

As shown in *Figure 14b*, the luminescence signal will slowly rise as ADP is phosphorylated to ATP by the pyruvate kinase system. When the signal stabilises at its maximum (after ~10 min), calibrate by adding ATP as above. the increase in signal due to the calibration is usually smaller due to the long incubation with pyruvate kinase (cf. *Figure 14a* and *b*). Although the method just given illustrates the principles involved, the analysis of ADP is actually best done by pre-treating the neutralised sample with pyruvate kinase and phosphenol pyruvate prior to sampling and ATP analysis.

(iv) *Calculations.* First the ATP content of the neutralised sample is calculated. From the calibration, calculate the increment in signal that is proportional to 1 nmol of added ATP. Then calculate the ATP content in the 100 μl of added sample after subtracting the signal seen before the sample addition from that observed after sample addition.

Repeat the calculation for the sample treated with pyruvate kinase. In this case, the ATP content found is due both to the ATP originally present and to ATP formed from the ADP content of the sample. To obtain the latter, subtract the original ATP content determined above. Some typical results are given in the legend to *Figure 14.*

(v) *Troubleshooting.* The assay is quantitative and reliable provided that the ATP concentration is below 1 μM. At higher concentrations, the accumulation of reaction products such as AMP becomes significant, resulting in inhibition of luciferase. This can lead to decay of the signal during the assay. To analyse high concentrations of ATP, it is necessary to dilute the sample so that the [ATP] in the assay is around 0.1 μM.

The method is usually trouble-free provided that precautions are taken to avoid contamination of glassware with ATP. This comes from improperly cleaned syringes and fungal or bacterial growth in the buffers and is only a problem because of the high sensitivity of the assay. Make sure that syringes used for ATP are cleaned using a weak, non-phosphate containing detergent. The Tris-acetate buffer must be made up freshly each day. Overnight storage in the refrigerator results in significantly higher background ATP readings, due presumably to the ATP content of organisms growing in the buffer.

(vi) *Upgrading.* This technique is easily adapted to measure ATP synthesis by thylakoid preparations. Use a medium such as AM5 (*Table 9*), but at a higher buffer concentration. It is important to include diadenosine pentaphosphate to inhibit chloroplast adenylate kinase, or else the added ADP will be partly converted to ATP and AMP in the dark pre-incubation period. Illuminate chloroplasts containing 20 μg of chlorophyll in 1 ml of AM5 for 1 min. Stop the reaction with 0.5 ml of 10% TCA. Neutralise the sample with 0.5 ml of 0.6 M KOH. At this point, the ATP content of the sample will be about 100 μM, which is too high for the direct addition of 100 μl to the luciferin assay. Therefore dilute the neutralised sample by at least 1:20 with Tris-acetate buffer. At this dilution, there is no need to spin the tubes to remove the precipitate and 100 μl samples can be taken directly for assay of ATP as described above.

4.3 Measurement of ATP Synthesis by Imposed Ion Gradients

This method epitomises the remarks made in the introduction to this chapter in that the technique for measuring ATP synthesis by imposed ion gradients is simple but the interpretation of the results is rather complex. The basic principle is to establish a transient $\Delta\bar{\mu}_{H^+}$ across the thylakoid in the absence of light by subjecting the membranes to a rapid acid/base transition, creating ΔpH (30). By manipulating the KCl concentrations and K^+ permeability (with valinomycin), a diffusion potential may also be transiently developed (31). Providing that the $\Delta\bar{\mu}_{H^+}$ initially created exceeds the thermodynamic and kinetic thresholds for ATP synthesis, some ATP will subsequently be made in the dark. The problem is that, once formed, $\Delta\bar{\mu}_{H^+}$ will immediately begin to decay as proton and other ion fluxes occur across the thylakoid membrane. Unless the experiment is performed in a stop-flow apparatus (17) so that ATP synthesis can be estimated at very short time intervals after the acid/base transition, net ATP synthesis will soon cease because $\Delta\bar{\mu}_{H^+}$ has become too small. The amount of ATP formed will thus depend on the amount of phosphorylating H^+ efflux that can be sustained after the acid/base transition. This in turn depends on a number of factors, such as the coupling state of the membrane, the initial magnitude of $\Delta\bar{\mu}_{H^+}$, the thermodynamic and kinetic thresholds for ATP synthesis, the internal buffering capacity of the intrathylakoid space and the electrical capacitance of the membrane.

(i) *Applications and advantages.* The method is useful for demonstrating the ability of thylakoids to synthesise ATP independently of light-driven electron transport (17,19,31). However, as discussed above, the amount of ATP formed cannot generally be used to calculate a rate of synthesis since $\Delta\bar{\mu}_{H^+}$ decays continuously. The method is most useful for studying the threshold $\Delta\bar{\mu}_{H^+}$ at which net ATP synthesis is first observed.

(ii) *Method.* The method described here is designed primarily to measure ATP synthesis in response to an acid/base transition (19). ATP synthesis is enhanced by simultaneously imposing a diffusion potential across the membrane, but if this needs to be precisely controlled and varied, then media other than the ones used here are more suitable (31). No special equipment is required other than that needed to analyse ATP synthesised by the ^{32}P or luciferin/luciferase methods described above.

The ATP synthesised during an acid/base transition will be hydrolysed in the following period if CF_0-CF_1 is in the thiol-reduced form (19). This can be avoided by using the ^{32}P method and including glucose/hexokinase to trap the newly incorporated ^{32}Pi. The following method is based on this principle and may be used for thylakoids in which CF_0-CF_1 is either thiol-oxidised or thiol-reduced.

(a) Prepare washed thylakoids as described in Section 2.2.
(b) Mix the stock suspension with thiol-modulation buffer LM2 (*Table 2*) to give thylakoids at a concentration of 0.6 mg chlorophyll ml^{-1}. If thiol-reduced CF_0-CF_1 is required, pre-illuminate as described in Section 2.2. Otherwise, omit DTT from LM2 and store on ice in the dark for no longer than 30 min.
(c) Take 100 μl of diluted thylakoids and mix with 0.1 ml of the acidic assay medium AM9 (*Table 12*).
(d) Pre-incubate in a stirred, disposable test tube (Section 2.2) for 30 sec, then rapidly add 0.8 ml of the basic buffer AM10 (*Table 12*).

Rapid mixing creates the transient ΔpH across the thylakoid and ATP synthesis will begin immediately but cease within 5 sec. The magnitude of the initial ΔpH created depends on the initial pH of both AM9 and AM10. If AM9 is at pH 4.5 and AM10 is at pH 8.5, a ΔpH of −4 units would be expected. In fact, the final pH after mixing is less than the initial pH of AM10, so ΔpH

Table 12. Media for the Assay of ATP Synthesised in Acid/Base Transitions.

AM9 (acidic buffer)	*AM10 (basic buffer)*
5 mM $MgCl_2$	5 mM $MgCl_2$
30 mM succinate-KOH	0.15 mM ADP
pH 4–6	2.5 μM diadenosine pentaphosphate
	15 mM glucose
	4 units hexokinase ml^{-1}
	0.15 mM NaH_2PO_4 containing 0.5 μCi ^{32}Pi ml^{-1}
	100 mM buffer
	pH 6.5–9.5

Prepare AM10 just prior to use. Suitable buffers for AM10 are MES, pH 6–7; Hepes, pH 7–8; Tricine, pH 7.5–8.5; glycylglycine pH 8–9.5.

will be slightly less than −4. For this reason, the final pH of the mixture should be checked in parallel experiments using a pH meter.

(e) 15 sec after mixing, add 0.5 ml of 5% TCA and assay the tubes for incorporated ^{32}Pi as described in Section 4.2.2. Calculate the ATP synthesised as nmol (mg chl)$^{-1}$.

Some of the inherent complexities in interpreting the results are evident from the data depicted in *Figure 15a*. In this experiment, the pH of the acidic buffer was kept constant at pH 5.4, whilst the pH of the basic buffer was varied in order to vary the ΔpH initially generated. Increasing the pH of the basic buffer from pH 6 to 9.5 progressively increases ΔpH from about −0.6 to −3.8, yet the yield of ATP synthesised did not progressively increase but passed through a maximum at about pH 8.8 (ΔpH = −3.4). This of course indicates that the yield of ATP formed depends not only on ΔpH but also on whether the external pH is optimal for the enzyme. Since the pH optimum for CF_0-CF_1 has generally been observed to be pH 8.0, the apparent optimum of pH 8.8 observed in this experiment represents a compromise between an external pH optimum of 8.0 and the higher values needed to generate a large ΔpH. It is clear that varying the pH of the basic medium introduces two variables affecting the yield of ATP synthesised after the acid/base transition. The situation is not improved by keeping the basic stage constant and varying the pH of the acid bath. In this case, there are still at least two variables because a decrease in internal pH would increase both ΔpH and the total buffering capacity of the intrathylakoid space. There is a third complication in that pH values in the acid bath below pH 5 cause irreversible inactivation of thylakoids which has to be corrected for by a separate assay of thylakoid function.

It is clear that a change in the yield of ATP observed on changing ΔpH is not simply due to ΔpH, but also to other variables that are unavoidably changing. These other variables become less important if one simply considers the threshold ΔpH needed to observe ATP synthesis. At this point, the value of ΔpH will reflect either the thermodynamic equilibrium of ATP synthesis or the kinetic activation of CF_0-CF_1 (see Section 3.1). *Figure 15a* shows that thiol-reduction of the ATPase lowers the threshold ΔpH required to observe net ATP synthesis (19). This indicates that the higher threshold seen when CF_0-CF_1 is thiol-oxidised must reflect an activation of the enzyme by ΔpH. It is currently thought that only the lower threshold ΔpH seen when the enzyme is thiol-reduced reflects the thermodynamic equilibrium, when the free energy change of ATP

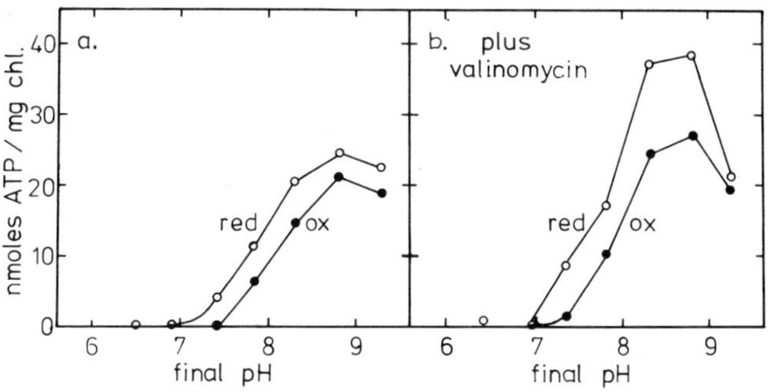

Figure 15. Yield of ATP synthesised after subjecting thylakoids to acid/base transitions. The pH of the acid stage (AM9) was 5.4 throughout. The pH of the base stage was varied by setting the pH of AM10 to 6.0, 6.5, 7.0, 7.5, 8.0, 8.5, 9.0 or 9.5. The final pH after rapid mixing was lower than the initial pH of AM10 as indicated in the figure. Thylakoids were either thiol-reduced (open circles) or thiol-oxidised (closed circles). The experiment was performed either in the absence **(a)** or presence **(b)** of 1 μM valinomycin.

synthesis is just balanced by the energy available from $\Delta\bar{\mu}_{H^+}$.

Figure 15b shows the effects of adding valinomycin to the system in the absence of added KCl. It can be seen that the yield of ATP is stimulated and the threshold ΔpH lowered. This indicates that the acid/base procedure also generates a diffusion potential across the thylakoid under these conditions, as would be expected as the K^+ content of the acid stage is about 15 mM whilst that of the basic stage is around 50 mM. In the presence of valinomycin, a diffusion potential of around 25 mV would be expected, and this is consistent with the lowered threshold ΔpH observed.

To minimise the diffusion potential developed as a result of the acid/base procedure, chloroplasts should be treated with valinomycin and a large and equal KCl concentration added to both the acidic and basic buffers AM9 and AM10. Alternatively, K^+ levels in both media can be controlled during preparation so that the magnitude of the diffusion potential can be approximately calculated.

4.4 Measurement of the Activation State of CF_0-CF_1

To study the influence of the activation state of the chloroplast ATPase on the rate of catalysis, one ideally requires an independent and quantitative method for measuring the fraction of enzyme complexes that are active at any one time. There is evidence that a non-catalytic adenine nucleotide binding site on CF_1 may provide such a measure, at least under certain conditions. It has been shown that $\Delta\bar{\mu}_{H^+}$ induces a change in the binding affinity of this site for ADP, such that a tightly bound nucleotide becomes freely exchangeable with ADP in the medium (17,18). The release of bound nucleotide correlates with the appearance of catalytic activity, and seems therefore to correlate with the activation of CF_0-CF_1. Similarly, deactivation of the enzyme that occurs when $\Delta\bar{\mu}_{H^+}$ is collapsed correlates with the appearance of tightly bound adenine nucleotide on the enzyme. The kinetics of release and rebinding of nucleotide are complex and not entirely in accord with the simple picture given above. However, further discussion is not possible here. The reader should be aware that this is a very active area of research, and the reliability of the method given below for measuring the activation state has yet to be established.

Table 13. Media for Assaying Bound [^{14}C]ADP to Thylakoid CF_0-CF_1.

AM11	*Quench and wash media*
5 mM $MgCl_2$	Quench medium
25 mM KCl	5 mM $MgCl_2$
0.1 mM methyl viologen	50 mM KCl
2.5 μM diadenosine pentaphosphate	200 mM NH_4Cl
5 mM NaH_2PO_4	20 μM FCCP
30 units hexokinase ml^{-1}	5 mM ADP
20 μM ADP containing 0.75 μCi [^{14}C]ADP ml^{-1}	25 mM Tricine-KOH
50 mM glycylglycine	pH 8.0
pH 8.0	Wash medium as above but omit FCCP and ADP and reduce NH_4Cl to 50 mM

Prepare media just prior to use.

(i) *Applications and advantages.* At present, this is about the only quantitative method for measuring the activation state of CF_0-CF_1 independently of assaying catalytic activity. It involves measuring the fraction of enzyme complexes that carry a tightly bound [^{14}C]ADP under steady-state conditions. This is assumed to represent the fraction of enzymes that are inactive at the time of the measurement. The assay is essentially a sampling techique and does not give a continuous measure of the activation state of CF_0-CF_1. However, it is possible to perform the assay under conditions where steady-state catalytic activity can be measured as described below.

(ii) *Method.* No special equipment is required other than that routinely used in ^{14}C radioisotope studies. The method described here is useful for measuring the level of [^{14}C]ADP bound under conditions of steady-state ATP synthesis.

(a) If thiol-reduced CF_0-CF_1 is required, dilute the chloroplast stock suspension with thiol-modulation buffer, LM2, and pre-illuminate (see Section 2.2). Otherwise, omit DTT from LM2 and do not illuminate.

(b) Take 0.2 ml of diluted chloroplasts and mix with 0.3 ml of assay medium AM11 (*Table 13*) in a disposable test tube. AM11 contains [^{14}C]ADP in a medium suitable for measuring ATP synthesis by the ^{32}P technique (Section 4.2.3).

(c) Pre-illuminate the tubes for 30 sec at an intensity of at least 100 W m^{-2} then allow at least 2 min in darkness.

This procedure causes bound ADP to exchange for [^{14}C]ADP in the medium. At this point, the thylakoids may be separated from the suspension medium by centrifugation, producing a preparation uniformly labelled with tightly-bound [^{14}C]ADP. In fact, this separation is not necessary for many experiments and the level of labelled nucleotide that is tightly bound under steady-state conditions can be assessed by simply re-illuminating the thylakoids as follows.

(d) During the dark period following the pre-illumination, make any additions of uncouplers or inhibitors as required. After a suitable pre-incubation in darkness (not less than 2 min), re-illuminate for 1 min and quench the system by rapidly adding 0.5 ml of quench medium (*Table 13*).

This medium contains a high concentration of uncouplers and unlabelled ADP.

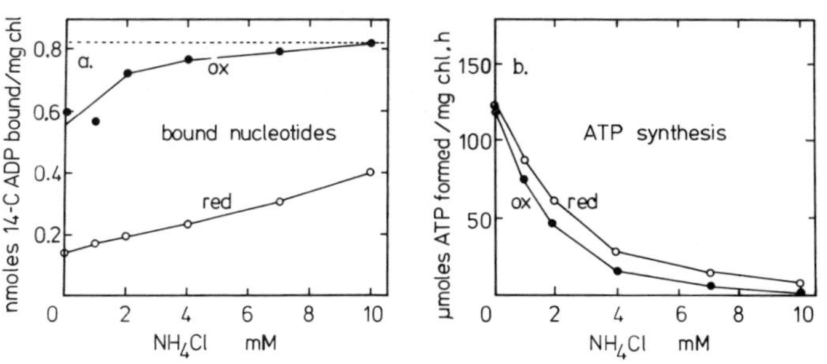

Figure 16. Comparison of **(a)** steady-state level of bound [^{14}C]ADP and **(b)** photophosphorylation measured by the ^{32}Pi method. Thylakoids were either thiol-reduced (open circles) or thiol-oxidised (closed circles). Rates of ATP synthesis are about 50% maximal because the ADP concentration is around the observed K_m (ADP) ($\sim$15 μM ADP at saturating light intensity).

The effect of its addition is to dissipate ΔpH rapidly and isotopically dilute the [^{14}C]ADP which is not tightly bound to CF_0-CF_1. Those enzyme molecules that are activated at the time of quenching will exchange their loosely bound [^{14}C]ADP for unlabelled ADP in the medium whilst those enzymes that are deactivated will retain the labelled ADP in a tightly bound form. Thus the amount of label that is retained by the thylakoids is a measure of the fraction of enzymes that are deactivated at the time of quenching.

(e) Leave the tubes for 10 min in the dark. As a control, pre-illuminate the thylakoids as above, allow 10 min in the dark, then add the quench medium. All the CF_0-CF_1 complexes in the control should retain labelled ADP.

(f) Pellet the membranes by centrifugation, remove the supernatant by aspiration and resuspend the pellet in 1 ml of wash medium (*Table 13*). Repeat the washing procedure twice to remove all traces of the labelled ADP in the suspension medium.

(g) The final pellet can be dissolved directly into a suitable scintillation cocktail (such as LKB 'Rialuma') and counted. However, the counts will be heavily quenched by the chlorophyll present.

To improve the counting efficiency, first resuspend the pellet in 1.0 ml of wash medium then precipitate the membranes by adding 0.5 ml of 10% TCA. The [^{14}C]ADP will be quantitatively released into the supernatant which can then be counted after removing the precipitate by centrifugation. In both cases, counting efficiencies need to be determined by internal standardisation.

(h) Count the vials, then add 0.1 ml of assay medium AM11 to several vials and recount to determine the c.p.m. per nmol of added [^{14}C]ADP.

The rate of ATP synthesis under these conditions can be determined by the ^{32}P technique as described in Section 4.2.3. Replace the labelled ADP in AM11 with unlabelled nucleotide and add ^{32}Pi.

Figure 16a shows some typical results obtained by the [^{14}C]ADP technique. CF_0-CF_1 was either thiol-oxidised or thiol-reduced and the rate of ATP synthesis was varied by adding NH_4Cl. *Figure 16b* shows the rate of ATP synthesis observed in parallel experiments using the ^{32}P technique. When the enzyme was thiol-oxidised, a high proportion of CF_0-CF_1 was inactive as indicated by the relatively high level of bound [^{14}C]ADP. This was increased even further by the addition of uncoupler over the concentration range that abolishes ATP synthesis. These results are consistent with the concept that, under these conditions, most ATPases are inactive and the rate of ATP synthesis is limited by the fraction of enzymes that are active. When CF_0-CF_1 was thiol-reduced, the [^{14}C]ADP bound in the light was much lower and relatively unresponsive to NH_4Cl even though ATP synthesis was inhibited as before. Taken at face value, the results suggest that enzyme activation no longer limits ATP synthesis and that, when CF_0-CF_1 is thiol-reduced, the rate of synthesis is controlled thermodynamically by ΔpH, as discussed in Section 3.1.

5. CONCLUDING REMARKS

Photophosphorylation is a complex process but the reactions of CF_0-CF_1 are technically easy to measure. The system can be approached at two levels. To demonstrate the process, relatively little apparatus is required and experiments can be done simply. ATP synthesis can be demonstrated with a pH meter (Section 4.2.2) and hydrolysis using a colorimetric assay (Section 4.2.1) and both methods are suitable for undergraduate teaching as well as research.

At a more advanced level, information on the magnitude of ΔpH and activation state of the enzyme are required in order to understand what controls the observed rate of catalysis. Using the apparatus described in Section 3.2, ΔpH can be determined in parallel, or even simultaneously with the rate of catalysis, whilst a method for measuring the activation state of CF_0-CF_1 has been given in Section 4.4. Combination of these techniques should allow the reader to explore the photophosphorylation system of thylakoids in a way demanded by the rigours of advanced research.

6. ACKNOWLEDGEMENTS

I would like to thank Professor R.E. McCarty for suggesting the labelled amine technique and Mr. G. Burgess for drawing the diagrams. Financial support for the experimental work was provided by the Science and Engineering Research Council and is gratefully acknowledged.

7. REFERENCES

1. Mitchell,P. (1968) *Chemiosmotic Coupling and Energy Coupling*, Available directly from Glynn Research Ltd., Bodmin, Cornwall PL30 4AU, UK.
2. Mitchell,P. (1977) *Eur. J. Biochem.*, **95**, 1.
3. Mills,J.D., Mitchell,P. and Schurmann,P. (1980) *FEBS Lett.*, **112**, 73.
4. Lowe,A.G. and Jones,M.N. (1984) *Trends Biochem. Sci.*, **9**, 11.
5. Witt,H.T. (1979) *Biochim. Biophys. Acta*, **505**, 355.
6. Davenport,J.W. and McCarty,R.E. (1980) *Biochim. Biophys. Acta*, **589**, 353.
7. Barber,J. (1972) *Biochim. Biophys. Acta*, **275**, 105.
8. Hind,G., Nakatani,H. and Izawa,S. (1974) *Proc. Natl. Acad. Sci. USA*, **71**, 1484.
9. Barber,J., Mills,J.D. and Love,A. (1977) *FEBS Lett.*, **74**, 174.
10. Vredenberg,W.J. (1976) in *The Intact Chloroplast: Topics in Photosynthesis*, Vol. **1**, Barber,J. (ed.), Elsevier, Amsterdam, p. 54.
11. Schuldiner,S., Rottenberg,H. and Avron,M. (1972) *Eur. J. Biochem.*, **25**, 64.
12. Portis,A.R. and McCarty,R.E. (1976) *J. Biol. Chem.*, **251**, 1610.
13. Tillberg,J.-E., Giersch,C. and Heber,U. (1977) *Biochim. Biophys. Acta*, **461**, 31.

14. Pick,U. and McCarty,R.E. (1980) in *Methods in Enzymology*, Vol. **69**, San Pietro,A. (ed.), Academic Press, New York, p. 539.
15. Fiolet,J.W.T., Bakker,E.P. and Van Dam,K. (1974) *Biochim. Biophys. Acta*, **368**, 432.
16. Horton,P. (1983) *Proc. R. Soc. Lond. B.*, **217**, 405.
17. Schlodder,E., Graber,P. and Witt,H.T. (1982) in *Electron Transport and Photophosphorylation: Topics in Photosynthesis*, Vol. **4**, Barber,J. (ed.), Elsevier, Amsterdam, p. 105.
18. Strotmann,H., Bickel-Sandkotter,S. and Shoshan,V. (1979) *FEBS Lett.*, **101**, 316.
19. Mills,J.D. and Mitchell,P. (1982) *FEBS Lett.*, **144**, 63.
20. Mills,J.D. and Mitchell,P. (1984) *Biochim. Biophys. Acta*, **704**, 93.
21. Rottenberg,H. (1979) *Biochim. Biophys. Acta*, **549**, 225.
22. Arana,J.L. and Vallejos,R.H. (1982) *J. Biol. Chem.*, **257**, 1125.
23. Bakker-Grunwald,T. and Van Dam,K. (1974) *Biochim. Biophys. Acta*, **347**, 290.
24. McCarty,R.E. and Racker,E. (1967) *J. Biol. Chem.*, **242**, 3435.
25. Gould,J.-M. (1978) *FEBS Lett.*, **94**, 90.
26. Steele,J.A., Uchytil,T.F. and Durbin,R.D. (1978) *Biochim. Biophys. Acta*, **504**, 136.
27. Nishimura,M., Ito,T. and Chance,B. (1962) *Biochim. Biophys. Acta*, **59**, 177.
28. Mitchell,P., Moyle,J. and Mitchell,R. (1979) in *Methods in Enzymology*, Vol. **55**, Fleisher,S. and Packer,L. (eds.), Academic Press, New York, p. 627.
29. Schlodder,E. and Witt,H.T. (1979) *Biochim. Biophys. Acta*, **635**, 571.
30. Jagendorf,A.T. and Uribe,E. (1966) *Proc. Natl. Acad. Sci. USA*, **55**, 170.
31. Hangatter,R.P. and Good,N.E. (1982) *Biochim. Biophys. Acta*, **681**, 397.

APPENDIX I

The Measurement of Radiant Energy

M.F. HIPKINS

Radiant energy can be measured by using one of three systems: the photometric, radiometric and quantum systems.

1. PHOTOMETRIC

The photometric system was specifically developed to measure light detected by the human eye. The system only measures wavelengths that fall in the visible region of the spectrum ($\sim 380-720$ nm), and the wavelength sensitivity of photometric instruments is a bell-shaped curve, with maximal response at about 550 nm and a decreased response at wavelengths removed from this position (*Figure 1*). The wavelength sensitivity closely resembles that of the human eye, but does not remotely correspond to photosynthetically active radiation (PAR). Photometric instruments are not of any use in measuring radiation that excites photosynthesis.

Units: Lux (lumen m^{-2}).

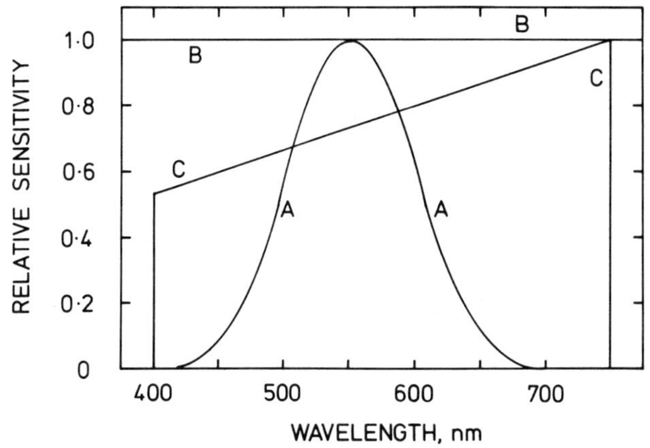

Figure 1. Diagrammatic representation of the ideal relative wavelength sensitivities of photometric **(A)**, radiometric **(B)** and quantum **(C)** detectors. Photometric sensitivity peaks at about 550 nm, and has very low values at the extremes of the visible spectrum. Radiometric detectors have a flat response, whilst quantum detectors compensate for the lower energy of long-wavelength photons relative to short-wavelength photons. Real wavelength sensitivities can differ substantially from the ideal: it is essential to check the specifications of the instrument for deviations from the responses shown here.

2. RADIOMETRIC

A radiometric detector is essentially a black-body absorber, which absorbs all radiation irrespective of its wavelength, and whose wavelength sensitivity is flat across the wavelength range of use (*Figure 1*). In the past, thermopiles were often used as the radiation detector, and their characteristics made them a good approximation to a black body. Nowadays it is more frequent to find semiconductor devices with special filters attached, and whose characteristics correspond to a black-body absorber over only a relatively restricted range of wavelengths, typically 380 – 850 nm. This covers the spectrum of PAR, and radiometric meters are often used to measure PAR.

The radiometric system uses units of energy, so light intensities (radiant fluence rates) are measured in W m^{-2} (J m^{-2} sec^{-1}). The disadvantage of radiometric measurements is that they convey no information about the number of photons in a beam of radiant energy. Such information is useful because photosynthesis is a photochemical process driven by photons, not by their aggregate energy.

3. QUANTUM FLUX

A quantum flux meter is designed so that it responds to the number of photons incident on its sensitive surface. Photons at the blue end of the visible spectrum have more energy than those at the red end: the wavelength sensitivity of a quantum flux detector is thus the inverse of the photon energy (*Figure 1*). Whilst quantum flux meters overcome the shortcomings of radiometers, the wavelength range of some instruments can be rather limited and, in particular, not extend much above 700 nm. Bearing this point in mind, quantum flux meters are the method of choice for measuring PAR.

Quantum flux densities (quantum fluence rates) are measured in μmol m^{-2} sec^{-1}. In some texts, the Einstein is used for a mole of photons, so an equivalent unit is μEinstein m^{-2} sec^{-1}.

APPENDIX II

Optical Filters

N.R. BAKER

Optical filters are used to modify or retard the passage of electromagnetic radiation. In photosynthesis studies the main types of filters used are interference, absorption, heat control and neutral density filters.

1. INTERFERENCE FILTERS

These filters rely upon optical interference to transmit relatively narrow bands of radiation; destructive interference is used to cancel out wavelengths of radiation not required for transmission, whilst constructive interference is used to reinforce the wavelengths to be transmitted. The typical characteristics of an interference filter are shown in *Figure 1*. Filters are characterised by the wavelength of the transmittance peak, the percentage of incident radiation transmitted at the peak and the bandwidth of the transmitted radiation. The bandwidth (sometimes termed half-power bandwidth) of a filter is defined as the width of the transmitted radiation band measured at 50% of the peak transmission (*Figure 1*). Decreasing the bandwidth (i.e. increasing spectral resolution) of a filter leads to a decrease in the amount of radiation transmitted. Interference filters can often transmit significant amounts of radiation at wavelengths distant from the maximum transmission wavelength, that is, at second order wavelengths. Generally, such second order transmissions are not required and can be prevented from reaching a sample by placing a suitable blocking absorption filter between the interference filter and the sample. Interference filters for transmissions throughout the u.v. and visible spectrum are readily available from many commercial outlets. Interference filters are expensive relative to absorption filters and should only be used when small, well-defined bandwidths of radiation are required.

2. ABSORPTION FILTERS

These filters are generally made of coloured glass or consist of a dye suspended in gelatin between glass plates. They absorb portions of the electromagnetic radiation spectrum and allow transmission of bands of radiation; the bandwidths of such filters generally range from about 30 to 250 nm and are considerably greater than those of interference filters. Absorption filters that have small bandwidths also absorb a large proportion of the radiation in the band required for transmission (*Figure 1*), such filters often have transmittances of less than 10% at the band peak.

Absorption filters can provide 'cut-off' characteristics, of which there are two types – long and short pass filters. Cut-off filters efficiently transmit over large regions of the spectrum but then rapidly decrease to a zero transmittance over a small wavelength

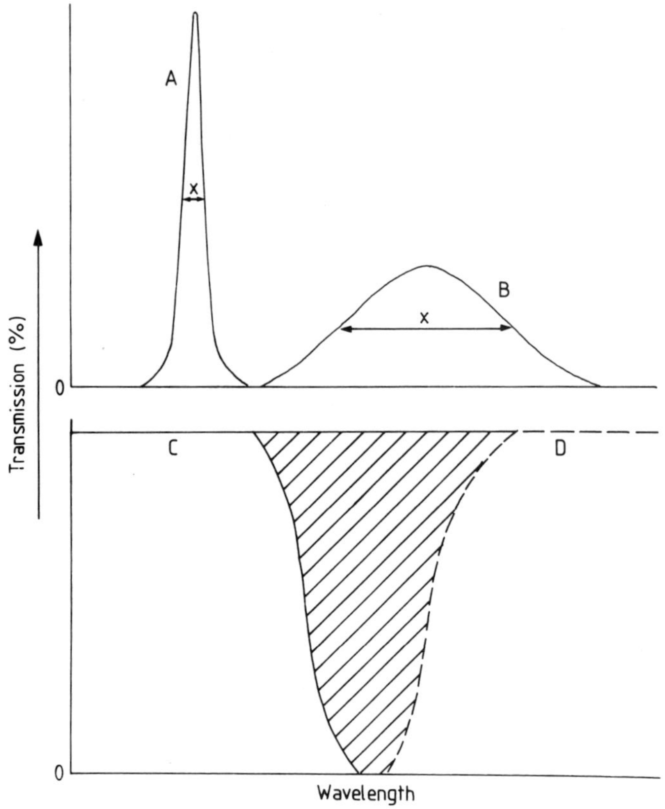

Figure 1. Transmission characteristics of **A**, an interference filter; **B**, an absorption filter; **C**, a short pass absorption filter; **D**, a long pass absorption filter. **x** designates the bandwidth of filters A and B.

change (*Figure 1*). Long pass filters transmit longer wavelengths whilst rejecting short wavelengths. Short pass filters transmit short wavelengths and reject longer wavelengths.

Absorption filters often have a tendency to bleach when exposed to high levels of light for long periods and consequently their transmission characteristics should be checked periodically with a good quality spectrophotometer. There are many commercial suppliers of coloured glass and gelatin filters.

3. HEAT CONTROL FILTERS

These filters transmit wavelengths below about 700 nm and reflect the higher, i.r. wavelengths. They are often termed heat-reflecting, or hot, mirrors and are used to protect interference and absorption filters and biological samples from the heat generated by light sources.

4. NEUTRAL DENSITY FILTERS

These filters are used to attenuate incident radiation without significantly modifying the spectral distribution of the transmitted radiation. Such filters are readily available with transmissions ranging from 80% to 0.01%. A major use of these filters is in the production of light (dosage) response curves for a given photosynthesis parameter. Neutral density filters do not always have the same transmittance at all wavelengths. Many filters do have a reasonably constant transmittance over the wavelength range 400 – 700 nm and these should be selected for photosynthesis studies using broad band light. As with absorption filters the transmission characteristics of these filters can change with prolonged exposure to high levels of light and thus they should be regularly checked.

INDEX

Published in the Practical Approach series

Plant cell culture

a practical approach

Edited by R A Dixon, *Royal Holloway College, London*

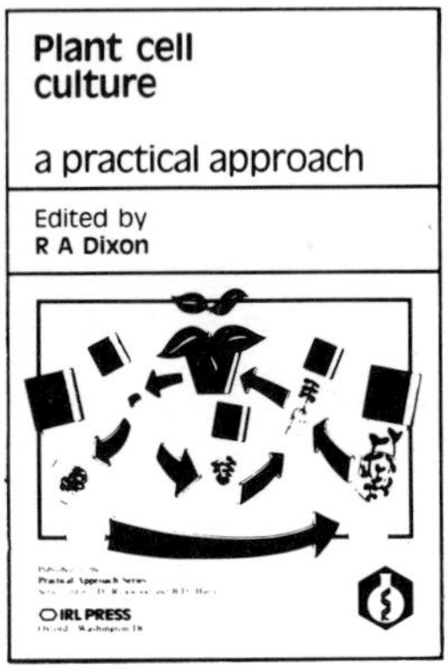

July 1985;
252pp;
0 947946 22 5
(softbound)

Different plant species respond differently to culturing and manipulation. *Plant cell culture* provides the knowledge necessary to rationalise these differing responses while, at the same time, offering effective laboratory-bench methods for dealing with particular cases.

Plant cells grown in aseptic culture offer exciting rewards for researchers in biotechnology and molecular biology. This new handbook departs from the agricultural emphasis of previous works to focus on plant cell culture as a successful and important tool for research.

The book describes how to establish cell cultures, the production and manipulation of protoplasts and the pathways of embryogenesis, organogenesis and regeneration. Other aspects described include vascular differentiation, selection for inhibitor resistance and the formation of secondary products. The book ends with plant – virus and plant – fungus interactions and protocols for cryopreservation of plant tissue.

Contents

Isolation and maintenance of callus and cell suspension cultures *R A Dixon* ● Haploid cell cultures *J M Dunwell* ● Isolation, culture and genetic manipulation of plant protoplasts *J B Power and J V Chapman* ● Selection of plant cells for desirable characteristics: inhibitor resistance *R A Gonzales and J M Widholm* ● Embryogenesis, organogenesis and plant regeneration *B Tisserat* ● Use of tissue cultures for studies on vascular differentiation *G P Bolwell* ● Secondary product formation by cell suspension cultures *P Morris, A H Scragg, N J Smart and A Stafford* ● Cryopreservation and storage of germplasm *L A Withers* ● Tissue culture methods in phytopathology I: viruses *K R Wood* ● Tissue culture methods in phytopathology II: fungi *S A Miller*

For details of price and ordering consult our current catalogue or contact:

IRL Press Ltd,
Box 1, Eynsham,
Oxford OX8 1JJ, UK

IRL Press Inc,
PO Box Q,
McLean VA 22101,
USA

IRL PRESS
Oxford · Washington DC